Big Data Analysis

Junwei Lu

Big Data Analysis

High Dimensional Probability, Statistics, Optimization, and Inference

Springer

Junwei Lu
Harvard University
Boston, MA, USA

ISBN 978-3-032-03160-0 ISBN 978-3-032-03161-7 (eBook)
https://doi.org/10.1007/978-3-032-03161-7

This Springer imprint is published by the registered company Springer Nature Switzerland AG
The registered company address is: Gewerbestrasse 11, 6330 Cham, Switzerland

To Professor Jianqing Fan's 60th birthday

Preface

The scope of this book is to serve as a concise yet comprehensive introduction to the mathematical and computational underpinnings of big data analysis. This book tries to provide the reader a quick yet comprehensive introduction, the necessary topics in modern data analysis spanning across probability, high-dimensional statistics, optimization, and inference.

This book is intended for graduate students, researchers, and professionals with a background in statistics, mathematics, computer science, or related fields who seek a rapid and robust introduction to high-dimensional data analysis. It is particularly suited for those eager to harness these methodologies for big data applications in academic or industrial settings. We aim to empower readers to quickly and effectively engage with the domain of high-dimensional statistics, equipping them with the conceptual and practical tools necessary for analyzing complex datasets. We hope this book could catalyze your journey into this vibrant and rapidly evolving field.

This book originates from the course BST235: Advanced Regression and Statistical Learning offered at Harvard T.H. Chan School of Public Health. I would like to thank my teaching fellows Shuting Shen and Zebin Wang for helping me with the course. I would also like to thank the students taking this course for their feedback and suggestions, which have been invaluable in refining the material in this book.

This book would not have been possible without the guidance, support, and encouragement of many individuals. I am deeply grateful to my parents for their unwavering belief in me and for their constant support throughout my life. In my journey to develop a comprehensive understanding of data science, I have been privileged to engage in insightful discussions and collaborations with exceptional colleagues from institutions worldwide. Their perspectives and feedback have been instrumental in shaping the ideas presented here. I would like to thank, among others, Han Liu, Jianqin Fan, Mladen Kolar, Philippe Rigollet, Tianxi Cai, Guang Cheng, Sébastien Bubeck, Mengdi Wang, Xihong Lin, Runze Li, Ming Yuan, Tony Cai, Hongzhe Li, Francis Bach, Peter Bickel, Lexin Li, Larry Wasserman, Richard Nickl, Richard Samworth, Ji Zhu, and Harry Zhou. They are heroes in my mind and my role models. I owe a debt of gratitude to the faculty, staff, and students at my home institution for fostering an enriching academic environment, as well as to the institutions that have hosted me during various stages of my career, providing invaluable support and inspiration.

The author has been partially supported by the National Science Foundation and National Institutes of Health. Lastly, I am thankful to the editorial and publishing teams who have guided this work to completion with their expertise and dedication. To all those whose names may have been inadvertently omitted, please know that your contributions are deeply appreciated. This work is a reflection of the collective efforts of a vibrant and supportive community.

Boston, MA, USA
January, 2025

Junwei Lu

Reading Guide

The book is structured into distinct yet interconnected chapters that reflect the core components of the data analysis pipeline:

Part I is an introduction to the topic of big data analysis. It also covers the review of the basic concepts in probability and statistics. Readers who are familiar with these concepts can skip this part.

Part II is about the high-dimensional probability, which is crucial for understanding the non-asymptotic behavior of random observations in high-dimensional settings. It mainly covers the concentration inequalities for independent random variables and martingales. It also covers maximal inequalities.

Part III is about the high-dimensional statistical methods and theory. It mainly focuses on the high-dimensional linear model with noiseless and noisy observations. It also introduces the generalized high-dimensional methods in other statistical problems. The method part of this part is independent of Part I, but the theory part is built on the concentration inequalities in Part I.

Part IV is about the high-dimensional optimization. We will cover the convex optimization and the gradient-based algorithms. It is independent of Parts I and II, but the optimization problems are motivated by the statistical problems in Part III.

Part V is about the high-dimensional inference, which is methods for hypothesis testing, confidence intervals, and variable selection in high-dimensional settings. It is built on the Part II of high-dimensional statistical methods.

This book refers to many other great textbooks and lecture notes. For the readers who are interested in specific topics and want to have a deeper understanding, I recommend the following books, which the book has benefited a lot from.
For high-dimensional probability:

- *High-Dimensional Probability: An Introduction with Applications in Data Science* by Roman Vershynin
- *Concentration Inequalities: A Nonasymptotic Theory of Independence* by Stéphane Boucheron, Gábor Lugosi, and Pascal Massart
- *Probability in High Dimension* by Ramon van Handel

For high-dimensional statistics:

- *High-Dimensional Statistics: A Non-asymptotic Viewpoint* by Martin J. Wainwright
- *Statistical Foundations of Data Science* by Jianqing Fan, Runze Li, Cun-Hui Zhang, and Hui Zou

For high-dimensional optimization:

- *Convex Optimization* by Stephen Boyd and Lieven Vandenberghe
- *Convex Optimization: Algorithms and Complexity* by Sébastien Bubeck

For high-dimensional inference:

- *Large-Scale Inference: Empirical Bayes Methods for Estimation, Testing, and Prediction* by Bradley Efron

Contents

Part I
Foundations of Big Data Analysis

Chapter 1
Introduction

1.1 What Is Big Data?

This book aims to solve two major questions:

1. *How to analyze big data? (Method)*
2. *Why it works? (Theory)*

To clarify the questions above, we need to define what is "big data." In this book, big data is almost a synonym of "high-dimensional data." The dataset, usually denoted as $\mathbb{X}$, is an $n \times d$ matrix, where n is the sample size and d is the number of features (or feature dimension). The following properties of big data are summarized:

Big data has three important features.

(1) **Volume**: both the sample size n and feature dimension d are both very large.
(2) **Velocity**: the scale of the data is so large such that the computation efficiency is crucial.
(3) **Variety**: the datasets are heterogeneous and of various types.

When talking about **high-dimensional statistics**, we typically emphasize the first feature and the feature dimension d is usually even (much) larger than the sample size n.

There are many tasks when analyzing big data. A typical protocol of data analysis is as follows:

$$\begin{array}{cccccccc} \text{Data} & \Longrightarrow & \text{Models} & \Longrightarrow & \text{Algorithms} & \Longrightarrow & \text{Decision}-\text{Making} \\ \uparrow & & \uparrow & & \uparrow & & \uparrow \\ \text{Probability} & & \text{StatisticalLearning} & & \text{Optimization} & & \text{Inference.} \end{array}$$

J. Lu, *Big Data Analysis*, https://doi.org/10.1007/978-3-032-03161-7_1

When analyzing data, of course we will start with datasets. Every statistician, as most people when viewing ratings on Amazon, has the intuition that the more samples we have in the dataset, the better information we will have. This intuition is supported by the probability theory by assuming the data is generated from specific statistical or machine learning models. In the big data analysis, both the sample size and feature dimensions are much larger than the classical analysis. We need to build high-dimensional statistical models. And when the model is specified, we need concrete algorithms to solve the model. Since the datasets are of large scale, optimization theory on how to develop fast algorithms is needed. Finally, we need to make decisions like rejecting the hypotheses and assessing the quality of our estimation. Statistical inference theory on the confidence interval and hypotheses testing is needed.

1.2 Three Principles

Seeing the typical protocol of data analysis above, we can see the four major cornerstones of modern big data analysis: probability, statistical learning, optimization, and inference, which therefore will be the four main topics of our book. Before diving into the details of the four topics, we would like to first talk about some philosophy. There are three major principles we will use multiple times in our book. They may seem vague the first time you see them, but we promise that we will make it rigorous when we implement these principles in different scenarios of big data analysis.

Concentration Principle: Random observations will converge to the population truth when the sample size becomes larger.

Immediate examples of concentration principle are the law of large numbers and the central limit theory. They are usually used in the theory of estimation and inference. However, to handle high-dimensional data, there are much richer probability theory developed beyond the law of large numbers and the central limit theory, which will be the major topic in the next chapter.

Parsimonious Principle: Although the feature dimension d is large, only a small proportion of features really play a role.

This principle plays a key role in high-dimensional statistics. It states that the high-dimensional datasets are only seemingly high dimensional. There is an essentially low-dimensional structured inside! It is a statistical version of the "Occam's razor": if two explanations are equally good, we prefer the simpler one. If two statistical models have a similar prediction accuracy, we always prefer the simpler and more interpretable one. This principle is also useful to avoid overfitting in data analysis. The word "simple" in the parsimonious principle has different meanings under different statistical models. We list two examples below on what do we mean by simpler models.

Example 1.1 (Sparsity) Consider the linear model

$$Y = \sum_{j=1}^{d} X_j \beta_j + \varepsilon.$$

We impose the sparsity assumption that only s of β_j's are nonzero and $s \ll d$.

The sparse linear model above is the most straightforward example of the parsimonious principle. Instead of considering a d-dimensional model, the sparse linear model only considers $\binom{d}{s}$ s-dimensional models. However, the parsimonious principle is not restricted to sparsity. It includes all regularization structures which make the model simple.

Example 1.2 (Additive Model) The additive model assumes that the general nonparametric model $Y = f(X_1, \ldots, X_d) + \varepsilon$ has the additive structure:

$$Y = \sum_{j=1}^{d} f_j(X_j) + \varepsilon.$$

The sparse additive model further assumes that only s of these d nonlinear functions $f_j(x)$ is not zero.

Taylor Principle: All functions (we are interested in) are "almost" quadratic.

The last principle gets its name by Taylor expansion. Although there are a variety of functions, in our lecture, we can find that most of the analysis is almost same as the quadratic functions. In statistical analysis, the Taylor principle implies that all (log)-likelihood functions are almost least squares. In optimization, we will introduce the concept of convexity and smoothness, which also states that the objective function is almost quadratic.

This principle plays a key role in high-dimensional statistics. It states that some high-dimensional datasets are only seemingly high dimensional. There is an essentially low-dimensional structure inside. It is a statistical version of the "Occam's razor": if two explanations are equally good, we prefer the simpler one. If two statistical models have a similar prediction accuracy, we always prefer the simpler and more interpretable one. This principle is also useful to avoid overfitting in data analysis. The word "simple" in the parsimony principle has different [illegible]

For example, [illegible]

$$\sum [illegible]$$

Chapter 2
Preliminaries in Probability

2.1 Basics of Probability

When we roll a die, the small cube with six different numbers on its six faces will never give us a determinist answer before it comes to rest. Einstein thought even "God does not throw dice." However, probabilists have to (as well as statisticians!). Actually, they have developed a dedicated language to describe the world of uncertainty. We will first begin with reviewing several fundamental terminologies in probability theory.

Definition 2.1 (Statistical Model) A *statistical model* $\mathcal{P}$ is a set of probability distributions indexed by a parameter θ. We denote $\mathcal{P} := \{p_\theta(x) \mid \theta \in \Theta\}$. The set Θ is called *a parameter space.*

- When the parameter space Θ is finite-dimensional, we call the statistical model as a **parametric model**.
- When the parameter space Θ is infinite-dimensional, we call the statistical model as a **nonparametric model**.

Definition 2.2 (Random Sample) The random variables $X_1, \ldots, X_n$ are called a *random sample* from $p_\theta(x)$ if $X_1, \ldots, X_n$ are independent and $X_i \sim p_\theta(x)$ for every $i = 1, 2, \ldots, n$. In other words, $X_1, \ldots, X_n$ are independent and identically distributed random variables. (We usually use the abbreviation i.i.d.)

Definition 2.3 (Data) In statistics, data $x_1, \ldots, x_n$ is the realization of the random sample $X_1, \ldots, X_n$.

People always get confused about these two concepts: random sample and data. A random variable is like a random number generator. For example, a die is a machine to generate the number from 1 to 6 with equal probability. Suppose you want to roll a die for n times, as the results of the n rollings are uncertain, you have a random sample $X_1, \ldots, X_n$. You can study its probability properties like

J. Lu, *Big Data Analysis*, https://doi.org/10.1007/978-3-032-03161-7_2

distribution functions, expectation, variance, etc. When you do the experiment to roll a die for n times and get n numbers $x_1, \ldots, x_n$, this n number is a realization of the random sample, namely, the data. Therefore, data are concrete numbers. It is meaningless to discuss the distribution of data as they are not random anymore.

Definition 2.4 (Distribution Functions) The *cumulative distribution function* (cdf) of a random variable X is defined as

$$F(x) := \mathbb{P}(X \leq x).$$

The *density function* or *probability density function* (pdf) $p(x)$ of a random variable X satisfies

$$p(x) = \frac{d}{dx} F(x).$$

For discrete random variables, the cumulative distribution functions are not differentiable. We define the *probability mass function* (pmf) $p(x)$ of a discrete random variable X as

$$p(x) = \mathbb{P}(X = x), \quad \text{for all } x \in X(\Omega).$$

We use $p_\theta(x)$ as a generic representation to represent a distribution function. It could be a cdf, pdf, or pmf. $X \sim p_\theta(x)$ means the random variable X has the distribution as $p_\theta(x)$.

Definition 2.5 (Expectation and Variance) If a continuous random variable X has the density function $p(x)$, the expectation and variance of X is defined as

$$\mathbb{E}(X) = \int_{-\infty}^{+\infty} x p(x) dx;$$

$$\text{Var}(X) = \mathbb{E}(X - \mathbb{E}(X))^2 = \int_{-\infty}^{+\infty} (x - \mathbb{E}(X))^2 p(x) dx.$$

When we have a random vector X, the covariance matrix of X is defined as

$$\Sigma := \text{Cov}(X) = \mathbb{E}(X - \mathbb{E}(X))(X - \mathbb{E}(X))^T.$$

Definition 2.6 (Statistic) Let $X_1, \ldots, X_n$ be a random sample and $T(x_1, \ldots, x_n)$ is a function from $\mathbb{R}^n$ to $\mathbb{R}$; we call the random variable $T(X_1, \ldots, X_n)$ a statistic. The distribution of the statistic $T(X_1, \ldots, X_n)$ is called the sampling distribution. When a statistic is an estimator of a parameter θ, we usually denote it as $\widehat{\theta}$.

Example 2.1 (Sample Mean and Covariance) The average of the random sample $X_1, \ldots, X_n$ is called sample mean, and we denote it by

$$\bar{X} = \frac{X_1 + \ldots + X_n}{n}.$$

The sample covariance of the random sample is defined by

$$\widehat{\Sigma} = \frac{1}{n}\sum_{i=1}^{n}(X_i - \bar{X})(X_i - \bar{X})^T.$$

2.2 Asymptotic Theory

We will never know which number it will turn out to be when we are rolling a die, but if we roll a fair die for sufficiently many times, we will find the average of all the results approaches $3.5 = (1 + 2 + 3 + 4 + 5 + 6)/6$. In general, when we get more and more data, the random sample will have some asymptotic properties.

In order to describe the asymptotic behavior of random variables, we should first define what is the limitation of random variables.

Definition 2.7 (Converge in Probability) A sequence of random variables $X_1, X_2, \ldots$ converges to μ in probability if

$$\lim_{n\to\infty} \mathbb{P}(|X_n - \mu| > \epsilon) = 0, \quad \text{for all } \epsilon > 0.$$

We denote it as $X_n \xrightarrow{P} X$.

Another important asymptotic property of random variable is to describe the limit distribution of random variables. We first define another kind of convergence of random variables.

Definition 2.8 (Converge in Distribution) A sequence of random variables $X_1, X_2, \ldots$ have cdf $F_1, F_2, \ldots$. $X_1, X_2, \ldots$ converges in distribution to a random variable X with cdf $F(x)$ if

$$\lim_{n\to\infty} F_n(x) = F(x)$$

at all x such that $F(x)$ is continuous. We denote it as

$$X_n \xrightarrow{d} F(x).$$

Definition 2.9 (Consistent Estimator) A statistic $\widehat{\theta}_n$ is a consistent estimator of θ if $\widehat{\theta}_n \xrightarrow{P} \theta$.

The consistency of an estimator is an example of how to describe the concentration principle by the probability language. However, it is not accurate enough as it does not have the information of how fast the estimator converges. Therefore, we introduce the O_P notation below.

Definition 2.10 (O_P Notations) Given a fixed sequence a_n and a sequence of random variables $X_1, X_2, \ldots$, we denote $X_n = o_P(a_n)$ if $X_n/a_n \xrightarrow{P} 0$.

We denote $X_n = O_P(a_n)$ if, for any $\epsilon > 0$, there exist $C > 0$ and $N > 0$ such that for all $n > N$,

$$\mathbb{P}(|X_n/a_n| > C) < \epsilon.$$

Notice that the definition of O_P is weaker than the argument that there exists a constant $C > 0$ such that $\lim_{n\to\infty} \mathbb{P}(|X_n/a_n| > C) = 0$ (think about why). However, the later definition is easier to use in practice.

Definition 2.11 (Statistical Rate) We say an estimator $\widehat{\theta}$ has the statistical rate $O(a_n)$ if $|\widehat{\theta} - \theta| = O_P(a_n)$.

Chapter 3
Preliminaries in Linear Algebra

3.1 Basics of Linear Algebra

Linearity is the simplest structure in mathematics. Let's review the basic notations and terminologies in linear algebra.

A is an $m \times n$ real matrix, written $A \in \mathbb{R}^{m\times n}$, if

$$A = \begin{bmatrix} a_{11} & a_{12} & \cdots & a_{1n} \\ a_{21} & a_{22} & \cdots & a_{2n} \\ \vdots & & & \vdots \\ a_{m1} & a_{m2} & \cdots & a_{mn} \end{bmatrix}$$

where $a_{ij} \in \mathbb{R}$. The (i, j)th entry of A is $A_{ij} = a_{ij}$.

The *transpose* of $A \in \mathbb{R}^{m\times n}$ is defined as

$$A^T = \begin{bmatrix} A_{11} & A_{21} & \cdots & A_{m1} \\ A_{12} & A_{22} & \cdots & A_{m2} \\ \vdots & & & \vdots \\ A_{1n} & A_{2n} & \cdots & A_{mn} \end{bmatrix} \in \mathbb{R}^{n\times m}.$$

In other words, $(A^T)_{ij} = A_{ji}$. Note: $\boldsymbol{x} \in \mathbb{R}^n$ is considered to be a column vector in $\mathbb{R}^{n\times 1}$.

Definition 3.1 (Sums and Products of Matrices) The sum of matrices $A \in \mathbb{R}^{m\times n}$ and $B \in \mathbb{R}^{m\times n}$ is the matrix $A + B \in \mathbb{R}^{m\times n}$ such that

$$(A + B)_{ij} = A_{ij} + B_{ij}.$$

J. Lu, *Big Data Analysis*, https://doi.org/10.1007/978-3-032-03161-7_3

The product of matrices $A \in \mathbb{R}^{m\times n}$ and $B \in \mathbb{R}^{n\times \ell}$ is the matrix $AB \in \mathbb{R}^{m\times \ell}$ such that

$$(AB)_{ij} = \sum_{k=1}^{n} A_{ik} B_{kj}.$$

The $n \times n$ *identity matrix*, denoted I_n or I for short, is

$$I = I_n = \begin{bmatrix} 1 & 0 & \cdots & 0 \\ 0 & 1 & \cdots & 0 \\ \vdots & & & \vdots \\ 0 & 0 & \cdots & 1 \end{bmatrix} \in \mathbb{R}^{n\times n}.$$

We have $IA = A = AI$.

If it exists, the *inverse* of A, denoted A^{-1}, is a matrix such that $A^{-1}A = I$ and $AA^{-1} = I$. If A^{-1} exists, we say that A is *invertible*. We have $(A^{-1})^T = (A^T)^{-1}$ and $(AB)^{-1} = B^{-1}A^{-1}$. The *trace* of a square matrix $A \in \mathbb{R}^{n\times n}$, denoted $\mathrm{tr}(A)$, is defined as $\mathrm{tr}(A) = \sum_{i=1}^{n} A_{ii}$. We have $\mathrm{tr}(AB) = \mathrm{tr}(BA)$ if AB is a square matrix. We say A is *symmetric* if $A = A^T$. The rank of a matrix A is the dimension of A's column space.

3.2 Eigenvalues and Eigenvectors

Given two vectors $u, v \in \mathbb{R}^d$, the *inner product* is $\langle u, v\rangle := u^T v = \sum_{j=1}^{d} u_j v_j$. We define the ℓ_2-norm of $\mathbf{u}$ as $\|u\| = \sqrt{|\langle u, u\rangle|}$. The cosine of the angle between u and v is $\cos\alpha = \frac{\langle u,v\rangle}{\|u\|\|v\|}$. We say u is orthogonal to v, denoted as $u \perp v$ if $\langle u, v\rangle = 0$.

Definition 3.2 (Eigenvalues and Eigenvectors) We say λ and v are the eigenvalue and eigenvector of a matrix A if $Av = \lambda v$.

Theorem 3.1 (Eigenvalue Decomposition) *A symmetric matrix $A \in \mathbb{R}^{d\times d}$ has:*

- *Real eigenvalues: $\lambda_1 \geq \ldots \geq \lambda_d$.*
- *Orthonormal eigenvectors: $u_1, \ldots, u_d$ such that $\|u_i\| = 1$ and $u_i \perp u_j$ for all $1 \leq i < j \leq d$,*

such that we have the eigenvalue decomposition $A = \sum_{i=1}^{d} \lambda_i u_i u_i^T$.

Denote $U = [u_1, \ldots, u_d] \in \mathbb{R}^{d\times d}$ and

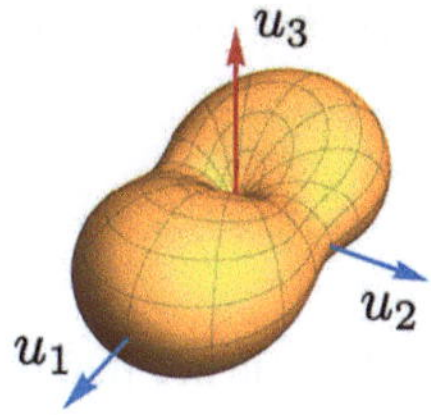

Fig. 3.1 Variational form of eigenvalues

$$\Lambda = \operatorname{diag}(\lambda_1, \ldots, \lambda_d) := \begin{bmatrix} \lambda_1 & 0 & \cdots & 0 \\ 0 & \lambda_2 & \cdots & 0 \\ \vdots & & & \vdots \\ 0 & 0 & \cdots & \lambda_d \end{bmatrix}.$$

The eigenvalue decomposition can also be written as $A = U\Lambda U^T$. We call U an *orthogonal matrix* as $U^T U = I_d$.

As a cornerstone of our chapter, we prefer to interpret the concepts as a solution of an optimization problem. Such interpretation is usually called variational form. The following theorem gives a variational form of eigenvalues.

Theorem 3.2 (Variational Form of Eigenvalues) *Given a symmetric matrix A the same as Theorem 3.1, its maximum eigenvalue has*

$$\lambda_{\max}(A) := \lambda_1 = \max_{\|x\|\le 1} x^T A x = \max_x \frac{x^T A x}{x^T x} \text{ and } u_1 = \operatorname*{argmax}_{\|x\|\le 1} x^T A x$$

and its minimum eigenvalue has

$$\lambda_{\min}(A) := \lambda_d = \min_{\|x\|\le 1} x^T A x = \min_x \frac{x^T A x}{x^T x} \text{ and } u_d = \operatorname*{argmin}_{\|x\|\le 1} x^T A x.$$

Question What is the variational form of other eigenvalues? See the visualization in Fig. 3.1.

The concept of eigenvalue decomposition can be generalized to nonsymmetric or even non-square matrices. We have the following theorem on singular value decomposition.

Theorem 3.3 (Singular Value Decomposition) *Given a rank r matrix $A \in \mathbb{R}^{n\times d}$, there exist:*

- *Singular values: $\sigma_1 \ge \ldots \ge \sigma_r > 0$ and denote $D = \operatorname{diag}(\sigma_1, \ldots, \sigma_r)$.*
- *Orthogonal matrices: $U = [u_1, \ldots, u_r] \in \mathbb{R}^{n\times r}$ and $V = [v_1, \ldots, v_r] \in \mathbb{R}^{d\times r}$ satisfying $U^T U = V^T V = I_r$,*

such that we have the singular value decomposition (SVD)

$$A = UDV^T = \sum_{i=1}^{r} \sigma_i u_i v_i^T.$$

We can see that the eigenvalue decomposition is a special SVD. The matrix A as a linear map transforms the two canonical unit vectors (yellow and red vectors) in the top left plot to the top right plot. This map can be decomposed into three steps: (1) V^T, rotating yellow and red vectors to v_1 and v_2; (2) D, scaling the disc by σ_1 horizontally and σ_2 vertically; and (3) U, rotating two canonical unit vectors (blue lines) to u_1 and u_2.

Part II
High-Dimensional Probability

Chapter 4
Concentration Inequalities

4.1 Asymptotic Versus Non-asymptotic

In the previous chapter, we discussed the concentration principle. It states that the more samples we have, the random observations converge to the population truth. In particular, we have the two important theorems in the probability theory describing this phenomenon.

Let $X_1, \ldots, X_n, \ldots$ be i.i.d. random variables with $\mathbb{E}(X_i) = \mu$ and $\text{Var}(X_i) = \sigma^2 < \infty$ for $i = 1, 2, \ldots$. Define the sample mean estimator $\bar{X}_n = (X_1 + \ldots + X_n)/n$. The first result describes the phenomenon that the average of random variables will converge to the expected value.

Theorem 4.1 (Law of Large Numbers) *The average $\bar{X}_n$ converges to mean, μ, in probability, i.e.,*

$$\bar{X}_n = \frac{1}{n}\sum_{i=1}^{n} X_i \xrightarrow{P} \mu.$$

Another important concentration property of random variable is to describe the limit distribution of the sample mean.

Theorem 4.2 (Central Limit Theorem) *The average $\bar{X}_n$ converges in distribution to the normal distribution. In particular, we have*

$$\sqrt{n}\left(\frac{\bar{X}_n - \mu}{\sigma}\right) \xrightarrow{d} N(0, 1).$$

Notice that both the law of large numbers and central limit theorem are the theorems characterizing the property of sample mean when the sample size n goes to infinity. We usually call this type of result the asymptotic property of random variables.

J. Lu, *Big Data Analysis*, https://doi.org/10.1007/978-3-032-03161-7_4

Although asymptotic properties like law of large numbers and central limit theorem tell us that some estimators converge to the truth, it has two major problems:

1. Asymptotic properties do not have the information of convergence rate.
2. Asymptotic properties may fail in high dimension when both n and d go to infinity.

The definition of convergence in probability or in distribution does not tell us how fast it converges. The law of large numbers implies that $\bar{X}_n$ is a consistent estimator while we do not know the statistical rate. The central limit theorem actually tells us the rate $|\bar{X}_n - \mu| = O_P(1/\sqrt{n})$. (Think about why.) However, this is only true when n goes to infinity. The second problem of the asymptotic properties is more severe. We would like to show why it may fail in high dimension by the following example.

Example 4.1 (Gaussian Mean Model) Let $X_1, \ldots, X_n$ be i.i.d. d-dimensional normal random vectors with the distribution $N(\mu, \sigma^2 I_d)$. We estimate the mean vector $\mu = (\mu_1, \ldots, \mu_d)^T$ by the sample mean $\bar{X}_n = \frac{1}{n}\sum_{i=1}^n X_i$. It is easy to check the sample mean is also normal, i.e., $\bar{X}_n \sim N(\mu, \sigma^2/n \cdot I_d)$. We measure the quality of the sample mean estimator by the mean squared error MSE $:= \mathbb{E}\|\bar{X}_n - \mu\|^2$. Using the covariance structure and the definition of ℓ_2-norm, we have

$$\mathbb{E}\|\bar{X}_n - \mu\|^2 = \frac{1}{n^2}\sum_{i=1}^{n}\sum_{j=1}^{d}\mathbb{E}(X_{ij} - \mu_j)^2 = \sigma^2\frac{d}{n}.$$

Since we have the closed form of the mean squared error, we can discuss it under different settings:

- When $d = 1$ or d is finite, we have MSE $= O(1/n)$ and the sample mean is a consistent estimator.
- When $d = \sqrt{n}$, we have MSE $= O(1/\sqrt{n})$ and the sample mean is still consistent but the rate becomes slower.
- When $d = n$, we have MSE $=$ constant and no longer converges as n goes to infinity.
- When $d = n^2$, MSE is diverging.

The example above shows that the asymptotic results may not be strong enough when we want to study the behavior of estimators when both d and n diverge. An idea to solve this problem is to first fix d and n as constants and see how the rate depends on them. We are interested in looking into the upper bound of tail probability like

$$\mathbb{P}(\|\bar{X}_n - \mu\|^2 > t) \le 2de^{-nt^2}$$

for any $t > 0$ and fixed d and n. We call this type of inequalities **non-asymptotic concentration inequalities** as it is true for all finite n and d.

4.2 Sub-Gaussian Random Variables

Let us start with the simplest tail probability inequality.

Theorem 4.3 (Markov Inequality) *Let X be a nonnegative random variable. For any $t > 0$, we have*

$$\mathbb{P}(X > t) \leq \frac{\mathbb{E}X}{t}.$$

Proof We prove the inequality for X when its pdf exists, and it is not hard to use a similar idea to prove the general case.

$$\mathbb{P}(X > t) = \int_t^{\infty} p(x)dx \leq \int_t^{\infty} \frac{x}{t} p(x)dx \leq \frac{1}{t} \int_{-\infty}^{\infty} xp(x)dx = \frac{\mathbb{E}X}{t}.$$

□

We can see that the Markov inequality shows that the tail probability decays at the rate $1/t$. Actually, we can achieve faster decaying rate by applying Markov inequality to $|X - \mathbb{E}[X]|^2$ and get the following **Chebyshev inequality**:

$$\mathbb{P}(|X - \mathbb{E}[X]| > t) = \mathbb{P}(|X - \mathbb{E}[X]|^2 > t^2) \leq \frac{\mathbb{E}|X - \mathbb{E}[X]|^2}{t^2} = \frac{\text{Var}(X)}{t^2}.$$

This gives us a $1/t^2$ decaying rate! You can probably imagine that we can actually further improve the rate by considering the kth moment:

$$\mathbb{P}(|X - \mathbb{E}[X]| > t) = \mathbb{P}(|X - \mathbb{E}[X]|^k > t^k) \leq \frac{\mathbb{E}|X - \mathbb{E}[X]|^k}{t^k}.$$

We now have a $1/t^k$ decaying rate for arbitrarily large k! Before we get too excited, we need to notice that there is a price to pay to get a faster and faster rate: the tail probability has the $1/t^k$ rate only if the kth moment of X is finite. Although it is not exactly a free lunch, the lunch is cheap enough: most random variables in reality are not that heavy-tailed.

So let us be aggressive again. Why only polynomial? Why not use the exponential function? This leads us to the so-called **Chernoff bound**:

$$\mathbb{P}(X - \mathbb{E}[X] > t) = \mathbb{P}(e^{\lambda(X - \mathbb{E}[X])} \geq e^{\lambda t}) \leq \frac{\mathbb{E}[e^{\lambda(X - \mathbb{E}[X])}]}{e^{\lambda t}}, \text{ for any } t \geq 0, \lambda \geq 0.$$

We now have an exponential decaying rate. Notice that we can also choose an optimal $\lambda > 0$ such that the right-hand side of the tail probability inequality is minimized. Let us formalize this idea in the following theorem.

Theorem 4.4 (Chernoff Bound) *Define the log-moment-generating function of X and its Legendre dual as*

$$\psi(\lambda) = \log \mathbb{E}[e^{\lambda(X-\mathbb{E}[X])}], \quad \psi^*(t) = \sup_{\lambda \geq 0}\{\lambda t - \psi(\lambda)\}.$$

We have $\mathbb{P}(X - \mathbb{E}[X] > t) \leq e^{-\psi^*(t)}$ *for all* $t \geq 0$.

Let's look at a concrete example for the Gaussian distribution.

Example 4.2 (Gaussian) Suppose $X \sim N(\mu, \sigma^2)$. We know that the moment-generating function for X is $\mathbb{E}[e^{\lambda(X-\mu)}] = e^{\frac{\lambda^2\sigma^2}{2}}$ and the log-moment-generating function is $\psi(\lambda) = \lambda^2\sigma^2/2$. We can get the closed form of its Legendre dual as

$$\psi^*(t) = \sup_{\lambda > 0}\left\{\lambda t - \frac{\lambda^2\sigma^2}{2}\right\} = \frac{t^2}{2\sigma^2} \text{ by choosing } \lambda = \frac{t}{\sigma^2}.$$

Then Chernoff bound gives us $\mathbb{P}(X-\mu > t) \leq \exp(-t^2/(2\sigma^2))$. The tail probability of Gaussian distribution is similar to its density function.

The Gaussian example above motivates us to consider the random variables with the tail probability similar to the Gaussian distribution.

Definition 4.1 (Sub-Gaussian) A random variable X is sub-Gaussian with the variance proxy σ^2, if

$$\mathbb{E}[e^{\lambda(X-\mathbb{E}[X])}] \leq e^{\frac{\lambda^2\sigma^2}{2}}, \quad \text{for all } \lambda \in \mathbb{R}.$$

Notice that the range of λ becomes the real line instead of only being positive in Theorem 4.4. This is because $\lambda \geq 0$ only gives us right-sided tail probability, and we will get the other side by considering $\lambda < 0$.

Theorem 4.5 (Two-Sided Tail Probability) *If X is sub-Gaussian with the variance proxy* σ^2, *we have*

$$\mathbb{P}(|X - \mathbb{E}[X]| > t) \leq 2e^{-\frac{t^2}{2\sigma^2}}.$$

Proof We get the right-sided tail probability $\mathbb{P}(X - \mathbb{E}[X] > t) \leq e^{-\frac{t^2}{2\sigma^2}}$ by directly using the Chernoff bound. We can bound the left-sided tail probability by considering $-X$:

$$\mathbb{P}(X - \mathbb{E}[X] < -t) = \mathbb{P}(-X - \mathbb{E}[-X] > t) \leq \frac{\mathbb{E}[e^{-\lambda(X-\mathbb{E}[X])}]}{e^{\lambda t}} \leq e^{-\lambda t + \frac{\lambda^2\sigma^2}{2}},$$

for all $\lambda > 0$. So we choose the optimal $\lambda = t/\sigma^2$ to minimize the tail probability and get $\mathbb{P}(X - \mathbb{E}[X] < -t) \leq e^{-\frac{t^2}{2\sigma^2}}$. We then have the two-sided tail probability bound by using union bound. □

From the proof of two-sided tail probability above, we find that the positive and negative domains of the moment-generating function imply the right- and left-sided tail probability, respectively. In this chapter, we sometimes only prove the one-sided tail probability of a random variable X and the other side can be proved similarly by considering $-X$.

Besides Gaussian random variables, our next example of sub-Gaussian random variable is the bounded random variable.

Theorem 4.6 (Bounded Random Variables) *If X is a bounded random variable, such that $a \leq X \leq b$ almost surely, for some $a, b \in \mathbb{R}$, then X is a sub-Gaussian with variance proxy $\frac{(b-a)^2}{4}$.*

Proof Intuitively, this theorem is obvious as the density of X goes directly to 0 when $t > b$, so it must have sub-Gaussian tail when t is large enough. However, the spirit of non-asymptotic concentration inequality is to prove the result for all $t > 0$ instead of letting t goes to infinity. So let us prove it by the definition of the sub-Gaussian.

Without loss of generality, we assume $\mathbb{E}[X] = 0$. Our goal is to find an upper bound of the log-moment-generating function $\psi(\lambda) = \log \mathbb{E}[e^{\lambda X}]$. Our idea is to study its first and second derivatives, as the Taylor principle tells us that we should consider its quadratic approximation. We have

$$\psi'(\lambda) = \frac{d}{d\lambda} \log \mathbb{E}[e^{\lambda X}] = \frac{\mathbb{E}[X e^{\lambda X}]}{\mathbb{E}[e^{\lambda X}]},$$

$$\psi''(\lambda) = \frac{d^2}{d\lambda^2} \log \mathbb{E}[e^{\lambda X}] = \frac{\mathbb{E}[X^2 e^{\lambda X}]}{\mathbb{E}[e^{\lambda X}]} - \left(\frac{\mathbb{E}[X e^{\lambda X}]}{\mathbb{E}[e^{\lambda X}]}\right)^2.$$

Define a new measure $dQ = \frac{e^{\lambda X}}{\mathbb{E}[e^{\lambda X}]} dP$, and we find that the derivatives of ψ become the expectation and variance under the new measure:

$$\psi'(\lambda) = \mathbb{E}_Q[X] \text{ and } \psi''(\lambda) = \text{Var}_Q(X).$$

We can bound the variance by

$$\psi''(\lambda) = \text{Var}_Q(X) = \text{Var}_Q\Big(X - \frac{a+b}{2}\Big) \leq \mathbb{E}_Q\Big[X - \frac{a+b}{2}\Big]^2 \leq \frac{(b-a)^2}{4},$$

where we use $a \leq X \leq b$ in the last inequality.

Notice that $\psi(0) = \log 1 = 0$ and $\psi'(0) = \mathbb{E}[X] = 0$. Using the fundamental theorem of calculus, we bound the log-moment-generating function by

$$\psi(\lambda) = \int_0^\lambda \int_0^\mu \psi''(p)dpd\mu \le \frac{\lambda^2(b-a)^2}{8}.$$

□

Notice the fact that $\psi''(\lambda) = \mathrm{Var}_Q(X)$ in the proof above is generally true for all random variables. This gives us an interesting insight: while Chebyshev inequality can only give a $1/t^2$-tail with a bound in variance, the Chernoff bound can give us a sub-Gaussian tail by also bounding the variance (everything is almost quadratic!) only under a different measure.

We show the concentration of sample mean of sub-Gaussian random variables.

Theorem 4.7 (Hoeffding Inequality) *Let $X_1, \ldots, X_n$ be independent random variables such that X_i is sub-Gaussian with variance proxy σ^2 for all $i = 1, \ldots, n$. We have*

$$\mathbb{P}\left(\frac{1}{n}\sum_{i=1}^n X_i - \mathbb{E}X > t\right) \le e^{-\frac{nt^2}{2\sigma^2}}.$$

Proof Again without loss of generality we assume $\mathbb{E}[X] = 0$. By Chernoff bound, we have

$$\mathbb{P}(\bar{X}_n > t) \le \mathbb{E}(e^{\lambda \sum_{i=1}^n X_i})e^{-n\lambda t} = \prod_{i=1}^n \mathbb{E}[e^{\lambda X_i}]e^{-n\lambda t} \le e^{n\left(\frac{\sigma^2\lambda^2}{2} - \lambda t\right)},$$

where the first equality is due to independence of X_i's and the last inequality is due to sub-Gaussian. So we finish the proof by choosing $\lambda = \frac{t}{\sigma^2}$. □

Similar to Theorem 4.5, we also have the two-sided result

$$\mathbb{P}\left(\left|\frac{1}{n}\sum_{i=1}^n X_i - \mathbb{E}X\right| > t\right) \le 2e^{-\frac{nt^2}{2\sigma^2}}.$$

We can write it in a high probability statement: with probability at least $1 - \delta$, we have

$$\left|\frac{1}{n}\sum_{i=1}^n X_i - \mathbb{E}X\right| \le \sqrt{\frac{2\sigma^2 \log(2/\delta)}{n}}, \text{ or } \left|\frac{1}{n}\sum_{i=1}^n X_i - \mathbb{E}X\right| = O_P(\sqrt{\sigma^2/n})$$

when we want to go back to the asymptotic statement.

Chapter 5
Sub-exponential Random Variables

5.1 Concentration Beyond Average

In the previous chapter, we showed the concentration of sample average in the Hoeffding inequality. The asymptotic results like law of large numbers and central limit theorem are also about the sample average. If we look into the proof of these results, we can find that these results rely on the additive formality of the sample mean. So we have the impression that the concentration principle works for the average, but does it cover other statistics? In fact, we have many nonlinear estimators in statistics and machine learning. Can we expect that a general statistic $f(X_1, \ldots, X_n)$ concentrates to its expectation? The answer is positive. Sample mean is not special. We have the general concentration principle as follows.

General Concentration Principle: A random variable $f(X_1, \ldots, X_n)$ concentrates to its mean $\mathbb{E}f(X_1, \ldots, X_n)$ if:

1. $X_1, \ldots, X_n$ are independent.
2. The function $f(x_1, \ldots, x_n)$ is not too "sensitive" to any of its coordinate x_i.

To make the above statement precise, we need to specify what do we mean by "sensitive"? One possible way is to measure it by coordinate derivatives. For example, for the sample mean, $f(x_1, \ldots, x_n) = \frac{1}{n}\sum_{i=1}^{n} x_i$. So the sample mean follows the general concentration principle because the derivative $\frac{\partial f}{\partial x_i} = \frac{1}{n}$, which goes to zero as n goes to infinity. Therefore, we can say the sample mean is not sensitive to any one of its observation. Can we use the derivatives in general to measure the sensitivity? The following theorem shows that the answer is positive, but we need to modify the definition of derivatives.

J. Lu, *Big Data Analysis*, https://doi.org/10.1007/978-3-032-03161-7_5

Theorem 5.1 (McDiarmid Inequality) *Define the ith discrete derivative of f*

$$D_i f = \sup_{x_1,\dots,x_n,x_i'} \Big| f(x_1,\dots,x_{i-1},x_i,x_{i+1},\dots,x_n) - f(x_1,\dots,x_{i-1},x_i',x_{i+1},\dots,x_n) \Big|.$$

For $X_1,\dots,X_n$ independent, we have for any $t>0$

$$\mathbb{P}\big(f(X_1,\dots,X_n) - \mathbb{E}f(X_1,\dots,X_n) > t\big) \le e^{-\frac{2t^2}{\sum_{i=1}^n (D_i f)^2}}.$$

Proof The main idea is to decompose the $f - \mathbb{E}f$ into additive form via the martingale method. Define the martingale difference

$$\Delta_k = \mathbb{E}[f(X_1,\dots,X_n)|X_1,\dots,X_k] - \mathbb{E}[f(X_1,\dots,X_n)|X_1,\dots X_{k-1}].$$

We then have the decomposition $f(X_1,\dots,X_n) - \mathbb{E}f(X_1,\dots,X_n) = \sum_{k=1}^n \Delta_k$. We call Δ_k as the martingale difference because $\mathbb{E}[\Delta_k|X_1,\dots,X_{k-1}] = 0$ by the tower property of conditional expectation.

Next, we will control the upper and lower bound of Δ_k. Define

$$U_k = \mathbb{E}\Big[\sup_z f(X_1,\dots,X_{k-1},z,X_{k+1},\dots X_n) - f(X_1,\dots,X_n)|X_1,\dots,X_{k-1}\Big],$$

$$L_k = \mathbb{E}\Big[\inf_z f(X_1,\dots,X_{k-1},z,X_{k+1},\dots X_n) - f(X_1,\dots,X_n)|X_1,\dots,X_{k-1}\Big]$$

and we have $L_k \le \Delta_k \le U_k$ almost surely. Notice that this is the only place we use the independence of $X_1,\dots,X_n$ throughout the proof.

As $|U_k - L_k| \le D_k f$, using Theorem 4.7, we have

$$\mathbb{E}[e^{\lambda\Delta_k}|X_1,\dots,X_{k-1}] \le e^{\lambda^2 (D_k f)^2/8}.$$

Notice that although we did not use $\mathbb{E}[\Delta_k|X_1,\dots,X_{k-1}] = 0$ explicitly, the above inequality is true only when Δ_k is a martingale difference.

We then compute

$$\mathbb{E}\big[e^{\lambda\sum_{k=1}^n \Delta_k}\big] = \mathbb{E}\Big[e^{\lambda\sum_{k=1}^{n-1}\Delta_k}\mathbb{E}\big[e^{\lambda\Delta_n}|X_1,\dots,X_{n-1}\big]\Big] \le \mathbb{E}\Big[e^{\lambda\sum_{k=1}^{n-1}\Delta_k} e^{\lambda^2(\Delta_n f)^2/8}\Big].$$

Repeating the above step inductively to $n-1,\dots,1$, we have

$$\mathbb{E}[e^{\lambda(f(X_1,\dots,X_n)-\mathbb{E}f(X_1,\dots,X_n))}] = \mathbb{E}\big[e^{\lambda\sum_{k=1}^n \Delta_k}\big] \le \prod_{k=1}^n e^{\lambda^2(D_k f)^2/8} = e^{\frac{\lambda^2}{8}\sum_{k=1}^n (D_k f)^2}.$$

So we complete the proof with the Chernoff bound of the moment-generating function above. □

Now, let's give a concrete example where McDiarmid inequality would be used.

Example 5.1 (Uniform Rate of Kernel Density Estimator) Let $X_1, \ldots, X_n$ be i.i.d. random variables from density $p(x)$ with support on $[0, 1]$. Our goal here is to estimate $p(x)$ by the kernel density estimator:

$$\widehat{p}(x) = \frac{1}{nh} \sum_{i=1}^{n} K\left(\frac{X_i - x}{h}\right),$$

where h is called the bandwidth and $K(\cdot)$ is the kernel function. Typically, we will choose the kernel function with bounded support satisfying

$$\int K(t)dt = 1, \int tK(t)dt = 0, \sup_z |K(z)| \le B \text{ and } |K(x) - K(y)| \le L|x - y|,$$

for some upper bound $B > 0$ and Lipschitz constant L. An example is the pyramid kernel $K(x) = \max(1 - |x|, 0)$. We aim to give the uniform statistical rate of $\widehat{p}$ via bounding

$$\|\widehat{p} - p\|_\infty := \sup_{x\in[0,1]} |\widehat{p}(x) - p(x)|.$$

We consider the variance-bias decomposition of $\|\widehat{p} - p\|_\infty$:

$$\text{Variance term} : \|\widehat{p} - \mathbb{E}[\widehat{p}]\|_\infty \text{ and Bias term} : \|\mathbb{E}[\widehat{p}] - p\|_\infty.$$

We first consider the variance term. Using the boundedness of the kernel function, we have

$$D_i(\|\widehat{p} - \mathbb{E}[\widehat{p}]\|_\infty) \le \frac{\|K\|_\infty}{nh} \le \frac{B}{nh}.$$

By McDiarmid inequality, we have, for any $t > 0$,

$$\mathbb{P}\Big(\big|\|\widehat{p} - \mathbb{E}[\widehat{p}]\|_\infty - \mathbb{E}\big[\|\widehat{p} - \mathbb{E}[\widehat{p}]\|_\infty\big]\big| > t\Big) \le 2e^{-2nh^2t^2/B^2}.$$

Therefore, with probability at least $1 - \delta$, we have

$$\big|\|\widehat{p} - \mathbb{E}[\widehat{p}]\|_\infty - \mathbb{E}\big[\|\widehat{p} - \mathbb{E}[\widehat{p}]\|_\infty\big]\big| \le \sqrt{B^2\frac{\log(2/\delta)}{2nh^2}}.$$

Now we bound the bias. The expectation of the kernel density estimator is

$$\begin{aligned}\mathbb{E}\widehat{p}(x) &= \int \frac{1}{h}K\left(\frac{t-x}{h}\right)p(t)dt \\ &= \int K(u)p(x+uh)du, \quad \text{(by change of variable } u=(t-x)/h) \\ &= \int K(u)(p(x)+p'(x)uh+p''(x)(uh)^2/2+O(h^3))du,\end{aligned}$$

where in the last equality we expand $p(x+uh)$ at x. So we can bound the bias as

$$\|\mathbb{E}[\widehat{p}]-p\|_\infty \le \frac{\|p''\|_\infty \int u^2K(u)du}{6}h^2+o(h^3)=O(h^2).$$

Let us summarize what we have so far:

$$\begin{aligned}\|\widehat{p}-p\|_\infty \le & \underbrace{\big|\|\widehat{p}-\mathbb{E}[\widehat{p}]\|_\infty - \mathbb{E}\big[\|\widehat{p}-\mathbb{E}[\widehat{p}]\|_\infty\big]\big|}_{\text{McDiarmid inequality: } O_P(1/\sqrt{nh^2})} + \underbrace{\mathbb{E}\big[\|\widehat{p}-\mathbb{E}[\widehat{p}]\|_\infty\big]}_{\text{Maximal: ??}} \\ & + \underbrace{\|\mathbb{E}[\widehat{p}]-p\|_\infty}_{\text{Bias: } O(h^2)}.\end{aligned}$$

We have bounded the first and third terms above. In the next lecture, we will show that the second term has the rate $O(\sqrt{\log n/(nh^2)})$. So we have

$$\|\widehat{p}-p\|_\infty = O_P\left(\sqrt{\frac{\log n}{nh^2}}+h^2\right).$$

Choosing the bandwidth $h \sim n^{-1/6}$, we have $\|\widehat{p}-p\|_\infty = O_P(\sqrt{\log n}\cdot n^{-1/3})$.

5.2 Sub-exponential Random Variables

In statistics, we usually need to estimate the variance or covariance matrix of a random variable X. Therefore, we need to consider X^2. However, even if X is sub-Gaussian, X^2 does not necessarily have a sub-Gaussian tail. The following example shows that even the square of Gaussian is not sub-Gaussian.

Example 5.2 (Gaussian) Let $X \sim N(0,1)$. The moment-generating function of X^2 is

$$\mathbb{E}e^{\lambda(X^2-1)} = \frac{1}{\sqrt{2\pi}}\int_{-\infty}^{\infty} e^{\lambda(z^2-1)}e^{-\frac{z^2}{2}}dz = \frac{e^{-\lambda}}{\sqrt{1-2\lambda}}.$$

We know that X^2 is not sub-Gaussian as its moment-generating function goes to ∞ when $\lambda \geq \frac{1}{2}$. However, if we take a closer look at the moment-generating function, we have

$$\frac{e^{-\lambda}}{\sqrt{1-2\lambda}} \leq e^{2\lambda^2}, \text{ as } |\lambda| < 1/2.$$

So if λ is not too large, the moment-generating function of X^2 is still bounded by something like sub-Gaussian. This motivates us for the definition of sub-exponential.

Definition 5.1 (Sub-exponential) We say a random variable X is sub-exponential with a parameter α if

$$\mathbb{E}\big[e^{\lambda(X-\mathbb{E}X)}\big] \leq e^{\frac{\lambda^2\alpha^2}{2}} \text{ for all } |\lambda| \leq \frac{1}{\alpha}.$$

The key difference between sub-Gaussian and sub-exponential is that for sub-Gaussian, the moment-generating function has the Chernoff bound for all $\lambda \in \mathbb{R}$, while for sub-exponential, it holds only for a specific range of λ. By the previous example, if $X \sim N(0, 1)$, X^2 is sub-exponential with parameter 2.

Now that we have a shorter range, what will happen to the tail probability of sub-exponential?

Theorem 5.2 (Sub-exponential Tail Probability) *If X is sub-exponential with a parameter α, then we have*

$$\mathbb{P}(X - \mathbb{E}X > t) \leq \begin{cases} e^{-\frac{t^2}{2\alpha^2}}, & \text{for } 0 \leq t < \alpha, \\ e^{-\frac{t}{2\alpha}}, & \text{for } t \geq \alpha. \end{cases}$$

Proof By Chernoff bound, for all $\lambda \in [0, \frac{1}{\alpha}]$, we have

$$\mathbb{P}(X - \mathbb{E}[X] > t) \leq \frac{\mathbb{E}e^{\lambda(X-\mathbb{E}[X])}}{e^{\lambda t}} \leq e^{-\lambda t + \frac{\lambda^2\alpha^2}{2}}.$$

We still want to choose the optimal λ minimizing the right-hand side above. If $\frac{t}{\alpha^2} \leq \frac{1}{\alpha}$, the optimal $\lambda = t/\alpha^2$ and we get the same result as sub-Gaussian: $e^{-\lambda t+\lambda^2\alpha^2/2} = e^{-t^2/(2\alpha)}$. If $\frac{t}{\alpha^2} > \frac{1}{\alpha}$, the optimal $\lambda = 1/\alpha$ and we have $e^{-\lambda t+\lambda^2\alpha^2/2} = e^{-t/\alpha+1/2} \leq e^{-t/(2\alpha)}$ as $1/2 < t/\alpha$. □

The sub-exponential tail probability has two types of range: it behaves like the sub-Gaussian when $0 \leq t < \alpha$, and when $t \geq \alpha$, the tail probability decays like e^{-t}.

Now we can study the sample average of the sub-exponential.

Theorem 5.3 *Let $X_1, \ldots, X_n$ be independent, $\mathbb{E}[X_i] = \mu$, and X_i is sub-exponential with a parameter α for all $1 \leq i \leq n$; we have*

$$\mathbb{P}(\bar{X}_n - \mu > t) \leq \begin{cases} e^{-\frac{nt^2}{2\alpha^2}} & \text{for } 0 \leq t \leq \alpha, \\ e^{-\frac{nt}{2\alpha}} & \text{for } t \geq \alpha. \end{cases}$$

Proof By Chernoff bound, for all $\lambda \in [0, \frac{1}{\alpha}]$, we have

$$\mathbb{P}(\bar{X}_n - \mu > t) \leq \frac{\mathbb{E}e^{\lambda(\bar{X}_n - \mu)}}{e^{\lambda t}} = \prod_{i=1}^{n} \mathbb{E}e^{\lambda(X_i - \mu)} e^{-\lambda nt} \leq e^{n(-\lambda t + \lambda^2 \alpha^2/2)}.$$

We can find that the right-hand side above is almost the same as the corresponding part of a single sub-exponential. We only have an additional n. So following the exact same analysis in the proof of Theorem 5.2, we can prove the result for sample mean. □

Chapter 6
Bernstein and Maximal Inequalities

6.1 Bernstein Inequality

In the previous chapter, we showed that the sample mean of independent sub-exponential random variables $X_1, \ldots, X_n$ with the parameter α has the tail probability

$$\mathbb{P}(|\bar{X}_n - \mathbb{E}X| > t) \leq= 2e^{-\frac{n}{2}\left(\frac{t^2}{\alpha^2} \wedge \frac{t}{\alpha}\right)},$$

where $x \wedge y = \min(x, y)$ and $x \vee y = \max(x, y)$. Therefore, with probability at least $1 - \delta$,

$$|\bar{X}_n - \mathbb{E}X| \leq \sqrt{\frac{\alpha^2}{n} \log\left(\frac{2}{\delta}\right) \vee \left(\frac{\alpha}{n} \log\left(\frac{2}{\delta}\right)\right)}.$$

We can see that the two types of sup-exponential tail probability give us two types of rate: $O(\alpha/\sqrt{n})$ and $O(\alpha/n)$. Although the second term is dominated by the first term, it implies the possibility of giving two types of rates in the concentration inequality. We are going to show a stronger concentration inequality of such type.

Definition 6.1 (Bernstein Condition) We say a random variable X with $\mathbb{E}X = 0$ and $\mathrm{Var}(X) = \sigma^2$ satisfies the Bernstein condition with parameter b if

$$\mathbb{E}|X|^k \leq \frac{\sigma^2}{2} k!\, b^{k-2}, \quad \text{for all } k \geq 3.$$

The Bernstein condition specified $k \geq 3$ because the $k = 1$ and $k = 2$ cases are vacuously true based on the existing assumptions about the mean and variance of X. It is straightforward to check that the bounded random variable satisfies the Bernstein condition.

J. Lu, *Big Data Analysis*, https://doi.org/10.1007/978-3-032-03161-7_6

Example 6.1 (Bounded Random Variable) If $|X| \le B$ and $\mathbb{E}X = 0$, we can check that X satisfies the Bernstein condition with parameter $B/3$ as follows:

$$\mathbb{E}|X|^k \le \mathbb{E}|X|^2 \cdot B^{k-2} \le \sigma^2 B^{k-2} = \sigma^2 \cdot 3^{k-2}\left(\frac{B}{3}\right)^{k-2} \le \sigma^2 \cdot \frac{k!}{2}\left(\frac{B}{3}\right)^{k-2}.$$

The last inequality is because $3^{k-2} \le \frac{k!}{2}$ for all $k \ge 3$.

In the following theorem, we will show that if X satisfies the Bernstein condition, then X is sub-exponential. We also have a better concentration inequality as follows.

Theorem 6.1 (Bernstein Inequality) *Let $X_1, \ldots, X_n$ be independent random variables with $\mathbb{E}X_i = 0$, $\mathrm{Var}(X_i) = \sigma^2$, and X_i satisfies the Bernstein condition with parameter b for all $i = 1, \ldots n$. We have*

$$\mathbb{P}(\bar{X}_n > t) \le \exp\left(-\frac{nt^2}{2(\sigma^2 + bt)}\right), \quad \textit{for all } t > 0.$$

Given a bounded random variable $|X| \le B$, Example 6.1 implies that X satisfies the Berstein condition with parameter $B/3$. Applying the Bernstein inequality, we have, with probability at least $1 - \delta$,

$$\bar{X}_n - \mathbb{E}X \le \frac{\sigma}{\sqrt{n}}\sqrt{\log(1/\delta)} + \frac{B}{3n}\log(1/\delta).$$

So Bernstein inequality gives us two types of rates like the rate of the sample mean of sub-exponential random variables. However, notice that the constant assigned to the $O(1/\sqrt{n})$ term is the variance σ^2. In contrast, if we apply the Hoeffding inequality, we have at least $1 - \delta$,

$$\bar{X}_n - \mathbb{E}X \le \frac{B}{\sqrt{n}}\sqrt{\log(1/\delta)}.$$

Therefore, Bernstein inequality will give us a better rate than the Hoeffding inequality when $\frac{\sigma}{\sqrt{n}} + \frac{B}{n} = o\left(\frac{B}{\sqrt{n}}\right)$, or equivalently, when $\sigma = o(B)$: the standard deviation is much smaller than the upper bound.

Proof We prove the Bernstein inequality by two steps. First, we show that if X satisfies the Bernstein condition, then X is sub-exponential. Second, we use the Chernoff bound to control the tail probability.

Step 1: We aim to prove that if X satisfies the Bernstein condition with parameter b, then X is sub-exponential with parameter $2(\sigma^2 \vee b)$.
We expand the moment-generating function of X into Taylor series:

$$
\begin{aligned}
\mathbb{E}e^{\lambda X} &= \sum_{k=0}^{\infty} \frac{\lambda^k}{k!}\mathbb{E}X^k = 1 + \frac{\lambda^2\sigma^2}{2} + \sum_{k=3}^{\infty}\frac{\lambda^k}{k!}\mathbb{E}X^k \\
&\le 1 + \frac{\lambda^2\sigma^2}{2} + \sum_{k=3}^{\infty}\frac{|\lambda|^k}{k!}\frac{\sigma^2}{2}k!b^{k-2} \qquad \text{(by the Bernstein condition)} \\
&= 1 + \frac{\lambda^2\sigma^2}{2}\sum_{k=0}^{\infty}|\lambda|^k b^k.
\end{aligned}
$$

If $|\lambda| < 1/b$, the infinite summation of the power series above converges, and we have

$$
\mathbb{E}e^{\lambda X} \le 1 + \frac{\lambda^2\sigma^2}{2}\sum_{k=0}^{\infty}|\lambda|^k b^k = 1 + \frac{\lambda^2\sigma^2}{2}\frac{1}{1-|\lambda|b} \le \exp\left(\frac{\lambda^2\sigma^2}{2(1-|\lambda|b)}\right),
$$

where we use $1+x \le e^x$ in the last inequality. Therefore, if $|\lambda| \le 1/(2(\sigma^2 \vee b))$, we further have

$$
\mathbb{E}e^{\lambda X} \le \exp\left(\frac{\lambda^2\sigma^2}{2(1-|\lambda|b)}\right) \le \exp(\lambda^2\sigma^2) \le \exp\left(\frac{\lambda^2}{2}\left(2(\sigma^2 \vee b)\right)\right).
$$

Therefore, by definition, X is sub-exponential with parameter $2(\sigma^2 \vee b)$.

Step 2: We apply the Chernoff bound on $\bar{X}_n$. By the analysis in Step 1, for all $|\lambda| < 1/b$, we have

$$
\mathbb{P}(\bar{X}_n > t) \le e^{-\lambda n t}\prod_{i=1}^{n} e^{\lambda X_i} \le \exp\left(\frac{n\lambda^2\sigma^2}{2(1-|\lambda|b)} - n\lambda t\right).
$$

We then complete the proof by choosing $\lambda = \frac{t}{\sigma^2+bt}$ as such $\lambda \in (0, 1/b)$. □

6.2 Maximal Inequality

In statistics, we sometimes need to study uniform performance of multiple estimators, especially in the high-dimensional scenario. In nonparametric statistics, we also need to study the uniform rate of some function estimator. Therefore, we need to study how large is the maximum of a set of random variables.

We start with the maximum of finite number of random variables. Given $X_1, \ldots, X_d$, we can control the maximum by $\mathbb{E}[\max_{1\le j\le d} X_j] \le \sum_{j=1}^{d}\mathbb{E}|X_j| = O(d)$. However, if X_i's are sub-Gaussian, we can achieve a much better upper bound.

Theorem 6.2 (Finite Maximal Inequality) *Let $X_1, \ldots, X_d$ be (not necessarily independent) sub-Gaussian random variables with variance proxy σ^2 and $\mathbb{E}X_i = 0$ for all $i = 1, \ldots, d$. We have*

$$\mathbb{E}\Big[\max_{1\le j\le d} X_j\Big] \le \sigma\sqrt{2\log d},$$

and with probability at least $1-\delta$,

$$\max_{1\le j\le d} X_j \le \sigma\sqrt{2\log(d/\delta)}.$$

Proof We will apply the Jensen's inequality that given a concave function φ, then $\varphi(\mathbb{E}X) \ge \mathbb{E}[\varphi(X)]$. See Fig. 6.1 for an illustration. So we rewrite the expectation as follows.

$$\begin{aligned}
&\mathbb{E}\Big[\max_{1\le j\le d} X_j\Big]\\
&= \mathbb{E}\Big[\frac{1}{\lambda}\log\exp\big(\max_{1\le j\le d}\lambda X_j\big)\Big]\\
&\le \frac{1}{\lambda}\log\mathbb{E}\Big[\exp\big(\max_{1\le j\le d}\lambda X_j\big)\Big] \quad \text{(by applying Jensen's inequality to } \log x\text{)}\\
&\le \frac{1}{\lambda}\log\sum_{j=1}^{d}\mathbb{E}e^{\lambda X_j} \le \frac{1}{\lambda}\log\sum_{j=1}^{d} e^{\frac{\lambda^2\sigma^2}{2}} = \frac{\log d}{\lambda} + \frac{\lambda\sigma^2}{2}.
\end{aligned}$$

We obtain the upper bound of expectation by choosing the optimal $\lambda = \sqrt{\frac{2}{\sigma^2}\log d}$.

Now we turn to control the tail probability:

$$\mathbb{P}\Big(\max_{1\le j\le d} X_j > t\Big) = \mathbb{P}\Big(\bigcup_{j=1}^{d}\{X_j > t\}\Big) \le \sum_{j=1}^{d}\mathbb{P}(X_j > t) \le de^{-\frac{t^2}{2\sigma^2}},$$

where the first inequality above is due to the union bound. □

Now we switch to study the maximal inequality for the suprema of infinite number of random variables. We first consider the ℓ_2-norm of a d-dimensional random vector $X = (X_1, \ldots, X_d)^T$ defined as $\|X\| = \sqrt{|X_1|^2 + \cdots + |X_d|^2}$. The ℓ_2-norm $\|X\|$ is a suprema due to its variational form

$$\|x\| = \sup_{\|y\|\le 1}\langle y, x\rangle,$$

which can be proved by using the Cauchy-Schwarz inequality $\langle y, x\rangle \le \|x\|\|y\|$ and the equality is obtained by choosing $y = x/\|x\|$.

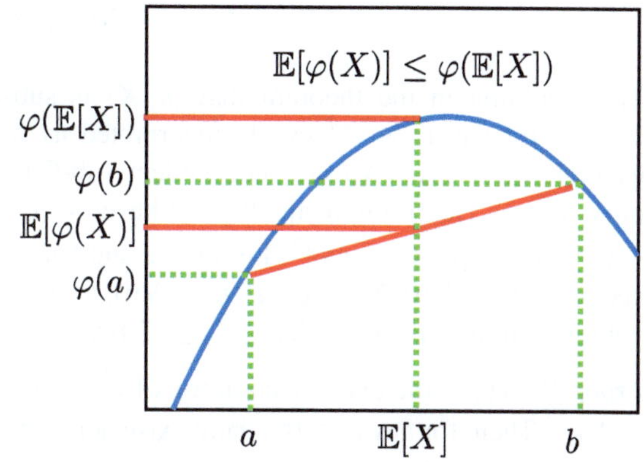

Fig. 6.1 Simple illustration of Jensen's inequality. For concave φ (blue line), $\varphi(\mathbb{E}X)$ will be greater than or equal to $\mathbb{E}[\varphi(X)]$

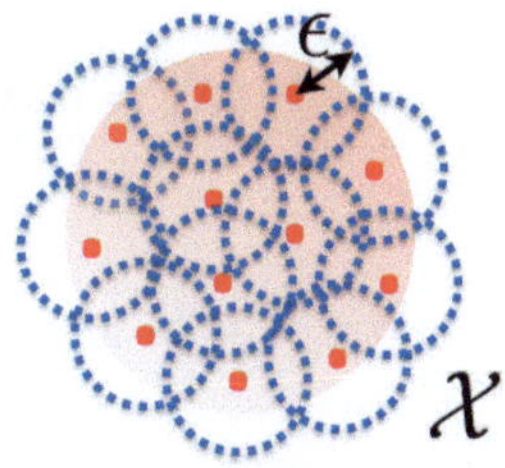

Fig. 6.2 Illustration of the ϵ-net $\mathcal{N}_\epsilon$ (red dots) of $\mathcal{X}$ (pink disc)

Denote the ℓ_2-ball as $\mathcal{B}_2^d = \{\theta \in \mathbb{R}^d | \|\theta\| \le 1\}$ and we can write $\|X\| = \sup_{u \in \mathcal{B}_2^d} \langle u, X \rangle$. Now we have a suprema over an ℓ_2-ball which is an infinite set. However, Theorem 6.2 only gives us the maximal inequality for finite random variables. Our idea to bridge the infinite set to a finite maximal inequality is the so-called discretization trick: we find finite representative points in the infinite set. This motivates us the following definition of ϵ-net.

Definition 6.2 (ϵ-Net) Given a metric space $\mathcal{X}$ with the distance function $d(\cdot, \cdot)$, we say $\mathcal{N}_\epsilon \subseteq \mathcal{X}$ is an ϵ-net of $\mathcal{X}$, if, for any $x \in \mathcal{X}$, there exists a $y \in \mathcal{N}_\epsilon$ such that $d(x, y) \le \epsilon$.

As illustrated in Fig. 6.2, an ϵ-net collects the centers of the balls with radius ϵ and covers the set $\mathcal{X}$. Therefore, we also define the minimum cardinality of all possible ϵ-net as the **covering number** of $\mathcal{X}$, denoted as $N(\mathcal{X}, d, \epsilon)$. Applying the discretization trick by considering the ϵ-net of the ℓ_2-ball as the representative points, we have the following result on the ℓ_2-norm of a random vector.

Theorem 6.3 (Maximal Inequality for ℓ_2-Norm) *Given a random vector $X \in \mathbb{R}^d$, if, for any $u \in \mathbb{R}^d$, we have $\langle u, X \rangle$ is sub-Gaussian with variance proxy $\sigma^2 \|u\|^2$, then we have*

$$\mathbb{E}\|X\| \le 4\sigma\sqrt{d}$$

and, with probability at least $1 - \delta$,

$$\|X\| \le 4\sigma\sqrt{d} + 2\sigma\sqrt{2\log(1/\delta)}.$$

The condition in the theorem that $\langle u, X\rangle$ is sub-Gaussian with variance proxy $\sigma^2\|u\|^2$ for any $u \in \mathbb{R}^d$ can be interpreted as X is a sub-Gaussian vector. For example, if $X_1, \ldots, X_d$ are independent sub-Gaussian scalars with the variance proxy σ^2, by Hoeffding inequality, we have $\langle u, X\rangle = \sum_{j=1}^d u_j X_j$ is sub-Gaussian with variance proxy $\sigma^2\|u\|^2$. Let us see another example for a dependent random vector. If $X \sim N(0, \Sigma)$, then $u^T X \sim N(0, u^T\Sigma u)$, and thus $u^T X$ is sub-Gaussian with the variance proxy $u^T\Sigma u \le \lambda_{\max}(\Sigma)\|u\|^2$.

Proof We apply the discretization trick by considering the 1/2-net of $\mathcal{B}_2^d$, denoted as $\mathcal{N}_{1/2}$. Then, for any $u \in \mathcal{B}_2^d$, there exists a $v \in \mathcal{N}_{1/2}$ such that $\|u - v\| \le 1/2$. So we can reformulate the ℓ_2-norm as:

$$\begin{aligned}\|X\| = \max_{u\in\mathcal{B}_2}\langle u, X\rangle &= \max_{u\in\mathcal{B}_2}\left(\langle v, X\rangle + \langle u - v, X\rangle\right)\\ &\le \max_{v\in\mathcal{N}_{1/2}}\langle v, X\rangle + \max_{\|u-v\|\le 1/2}\langle u - v, X\rangle\\ &\le \max_{v\in\mathcal{N}_{1/2}}\langle v, X\rangle + \frac{1}{2}\max_{u\in\mathcal{B}_2}\langle u, X\rangle.\end{aligned}$$

Rearranging the last inequality, we obtain the key inequality in the discretization trick:

$$\max_{u\in\mathcal{B}_2}\langle u, X\rangle \le 2\max_{v\in\mathcal{N}_{1/2}}\langle v, X\rangle,$$

which reduces a suprema over an infinite set to the maximum over a finite set. It follows from Theorem 6.2 that

$$\mathbb{E}\|X\| = \mathbb{E}\max_{u\in\mathcal{B}_2}\langle u, X\rangle \le 2\mathbb{E}\max_{v\in\mathcal{N}_{1/2}}\langle v, X\rangle \le 2\sigma\sqrt{2\log\left(|\mathcal{N}_{1/2}|\right)},$$

and

$$\mathbb{P}(\|X\| > t) = \mathbb{P}\Big(\max_{u\in\mathcal{B}_2}\langle u, X\rangle > t\Big) \le \mathbb{P}\Big(\max_{v\in\mathcal{N}_{1/2}}\langle v, X\rangle > t/2\Big) \le |\mathcal{N}_{1/2}|e^{-\frac{t^2}{8\sigma^2}}.$$

It remains to bound the covering number $|\mathcal{N}_{1/2}|$ by the following lemma.

Lemma 6.1 (Covering Number of ℓ_2-Ball) *Given any $\epsilon \in (0, 1)$, let $\mathcal{N}_\epsilon$ be the ϵ-net of $\mathcal{B}_2^d$. We have $|\mathcal{N}_\epsilon| \le \left(1 + \frac{2}{\epsilon}\right)^d$.*

Implementing the above lemma, we finish the proof of the theorem as

$$\mathbb{E}\|X\| \leq 2\sigma\sqrt{2\log\left(|\mathcal{N}_{1/2}|\right)} \leq 2\sigma\sqrt{2d\log 5} \leq 4\sigma\sqrt{d},$$

and the tail probability has $\mathbb{P}(\|X\| > t) \leq |\mathcal{N}_{1/2}|e^{-\frac{t^2}{8\sigma^2}} \leq 5^d e^{-\frac{t^2}{8\sigma^2}}$. Letting $\delta = 5^d e^{-\frac{t^2}{8\sigma^2}}$, we solve $t \leq 4\sigma\sqrt{d} + 2\sigma\sqrt{2\log(1/\delta)}$. Notice here we relax the constants to make them simplified. □

We now go back and prove Lemma 6.1.

Proof As the ϵ-net implies a covering (see Fig. 6.2), we have

$$\bigcup_{x\in\mathcal{N}_\epsilon}\{x+\epsilon\mathcal{B}_2^d\} \supseteq \mathcal{B}_2^d.$$

So the volume of an ℓ_2-ball is bounded by the volumes of all small balls, i.e.,

$$\mathrm{Vol}(\mathcal{B}_2^d) \leq \mathrm{Vol}\left(\bigcup_{x\in\mathcal{N}_\epsilon}\{x+\epsilon\mathcal{B}_2^d\}\right) \leq |\mathcal{N}_\epsilon|\mathrm{Vol}(\epsilon\mathcal{B}_2^d) = |\mathcal{N}_\epsilon|\epsilon^d\mathrm{Vol}(\mathcal{B}_2^d).$$

So this simple observation using the fact of covering gives us the lower bound of the covering number $|\mathcal{N}_\epsilon| \geq (1/\epsilon)^d$, which increases exponentially with the dimension. However, we need a finer analysis to show the upper bound, using a similar **volume argument**.

Let us construct the ϵ-net of $\mathcal{B}_2^d$ by a concrete procedure below:

1. Set $x_1 = 0$.
2. For $i \geq 2$, set $x_i = \mathcal{B}_2^d \setminus \bigcup_{j=1}^{i-1}\{x_j + \epsilon\mathcal{B}_2^d\}$. This is to say, pick a point in $\mathcal{B}_2$ not included in the union of the ϵ-balls around any of the existing x_i's.
3. Output: $\mathcal{N}_\epsilon = \{x_1, \ldots, x_{i-1}\}$ if there is no such x_i exists in Step 2.

This algorithm yields three important facts (see Fig. 6.3 for an illustration):

1. $\mathcal{N}_\epsilon$ is an ϵ-net, as we will keep adding points until we have no more space in Step 2.
2. The balls $\{x_j + \frac{\epsilon}{2}\mathcal{B}_2^d\}$ for all $x_j \in \mathcal{N}_\epsilon$ are disjoint, as based on this algorithm, $x_j, x_k \in \mathcal{N}_\epsilon$, $\|x_j - x_k\| > \epsilon$.

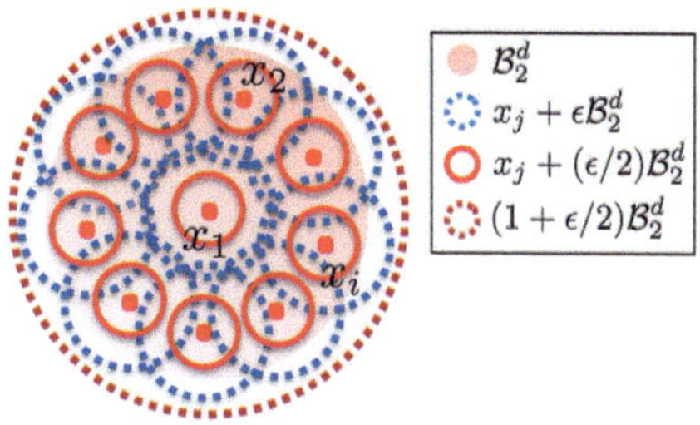

Fig. 6.3 Illustration of volume argument in the proof of Lemma 6.1

3. $\left(1+\frac{\epsilon}{2}\right)\mathcal{B}_2^d \supseteq \bigcup_{j=1}^{|\mathcal{N}_\epsilon|}\left\{x_j+\frac{\epsilon}{2}\mathcal{B}_2^d\right\}$, as $x_j \in \mathcal{B}_2^d$, and thus, the ball $\{x_j+\frac{\epsilon}{2}\mathcal{B}_2^d\}$ are contained in $\left(1+\frac{\epsilon}{2}\right)\mathcal{B}_2^d$.

Using these facts, we use the volume argument that

$$\begin{aligned}\left(1+\frac{\epsilon}{2}\right)^d \mathrm{Vol}(\mathcal{B}_2^d) &= \mathrm{Vol}\Big(\left(1+\frac{\epsilon}{2}\right)\mathcal{B}_2^d\Big)\\ &\geq \mathrm{Vol}\Big(\bigcup_{j=1}^{|\mathcal{N}_\epsilon|}\left\{x_j+\frac{\epsilon}{2}\mathcal{B}_2^d\right\}\Big) && \text{(by Fact 3.)}\\ &= \sum_{j=1}^{|\mathcal{N}_\epsilon|}\mathrm{Vol}\left(\frac{\epsilon}{2}\mathcal{B}_2^d\right) = |\mathcal{N}_\epsilon|\left(\frac{\epsilon}{2}\right)^d \mathrm{Vol}(\mathcal{B}_2^d). && \text{(by Fact 2.)}\end{aligned}$$

Solving the inequality above, we have $|\mathcal{N}_\epsilon| \leq \left(1+\frac{2}{\epsilon}\right)^d$. □

We finish this chapter by considering another example of implementing the discretization trick to maximal inequality.

Example 6.2 (Uniform Rate of the Kernel Density Estimator (Continued)) We discuss the kernel density estimator in Example 5.1. Given i.i.d. observations $X_1, \ldots, X_n$ generated from the density $p(x)$ supported on $[0, 1]$, recall the kernel density estimator:

$$\widehat{p}(x) = \frac{1}{nh}\sum_{i=1}^{n} K\left(\frac{X_i - x}{h}\right),$$

where the kernel $K(\cdot)$ is a bounded support function satisfying

$$\int K(t)dt = 1, \int tK(t)dt = 0, \sup_z |K(z)| \leq B \text{ and } |K(x) - K(y)| \leq L|x-y|,$$

for any x, y. We aim to bound the uniform rate $\|\widehat{p} - p\|_\infty := \sup_{x\in[0,1]} |\widehat{p}(x) - p(x)|$.

We can decompose $\|\widehat{p} - p\|_\infty$ into three terms:

$$\begin{aligned}\|\widehat{p} - p\|_\infty \leq &\underbrace{\Big|\|\widehat{p} - \mathbb{E}[\widehat{p}]\|_\infty - \mathbb{E}\big[\|\widehat{p} - \mathbb{E}[\widehat{p}]\|_\infty\big]\Big|}_{\text{Term I}} + \underbrace{\mathbb{E}\big[\|\widehat{p} - \mathbb{E}[\widehat{p}]\|_\infty\big]}_{\text{Term II}}\\ &+ \underbrace{\|\mathbb{E}[\widehat{p}] - p\|_\infty}_{\text{Term III}},\end{aligned}$$

where in Example 5.1, we have shown that Term I has the rate $O_P(1/\sqrt{nh^2})$ by McDiarmid inequality and the bias Term III has the rate $O(h^2)$. It remains to study

Term II, which is an expectation of maximal:

$$\begin{aligned}
\text{Term II} &= \mathbb{E}\big[\|\widehat{p} - \mathbb{E}[\widehat{p}]\|_\infty\big] \\
&= \mathbb{E}\Big[\sup_{x\in[0,1]} \big|\widehat{p}(x) - \mathbb{E}[\widehat{p}(x)]\big|\Big] \\
&= \mathbb{E}\Bigg[\sup_{x\in[0,1]} \bigg|\frac{1}{nh}\sum_{i=1}^{n} K\left(\frac{X_i - x}{h}\right) - \mathbb{E}K\left(\frac{X_i - x}{h}\right)\bigg|\Bigg].
\end{aligned}$$

To simplify our notation, we denote

$$K_i(x) = K\left(\frac{X_i - x}{h}\right) - \mathbb{E}K\left(\frac{X_i - x}{h}\right), \text{ for all } i = 1, \dots, n.$$

To control the maximal, we use the discretization trick and consider a $1/n$-net of $[0, 1]$. We can simply use the grids of the interval $\mathcal{N}_{1/n} = \{x_j = j/n | j = 1, \dots, n\}$. Similar to the proof of Theorem 6.3, we have for any $x \in [0, 1]$, there exists a $1 \le j \le n$ such that $|x - x_j| \le 1/n$ and thus

$$\begin{aligned}
\mathbb{E}\|\widehat{p} - \mathbb{E}\widehat{p}\|_\infty &= \frac{1}{nh}\mathbb{E}\Bigg[\sup_{x\in[0,1]} \bigg|\sum_{i=1}^{n} K_i(x)\bigg|\Bigg] \\
&\le \frac{1}{nh}\mathbb{E}\Bigg[\max_{1\le j\le n} \bigg|\sum_{i=1}^{n} K_i(x_j)\bigg| + \sup_{|x-x_j|\le 1/n} \sum_{i=1}^{n} |K_i(x) - K_i(x_j)|\Bigg].
\end{aligned} \tag{6.1}$$

For the first term in (6.1), it is a finite maximal. As $|K_i(x_j)| \le 2B$ for all $1 \le i, j \le n$, so by Hoeffding inequality, $\sum_{i=1}^{n} K_i(x_j)$ is sub-Gaussian with variance proxy nB^2. Applying the finite maximal inequality in Theorem 6.2, we have

$$\frac{1}{nh}\mathbb{E}\Bigg[\max_{1\le j\le n} \bigg|\sum_{i=1}^{n} K_i(x_j)\bigg|\Bigg] \le \sqrt{\frac{2B^2 \log(2n)}{nh^2}},$$

where we reduce the maximal of absolute values to the maximal inequality without absolute values via the identity $|X| = X \vee (-X)$.

For the second term in (6.1), we apply the Lipschitz property that $|K(x) - K(y)| \le L|x - y|$ for any x, y and have

$$\begin{aligned}
\frac{1}{nh}\mathbb{E}\Bigg[\sup_{|x-x_j|\le 1/n} \sum_{i=1}^{n} |K_i(x) - K_i(x_j)|\Bigg] &\le \frac{1}{nh}\mathbb{E}\Bigg[\sup_{|x-x_j|\le 1/n} \sum_{i=1}^{n} 2L\bigg|\frac{x - x_j}{h}\bigg|\Bigg] \\
&\le \frac{2L}{nh^2}.
\end{aligned}$$

Summarizing the rates of the two terms in (6.1), we have

$$\mathbb{E}\|\widehat{p} - \mathbb{E}\widehat{p}\|_\infty \le \sqrt{\frac{2B^2 \log(2n)}{nh^2}} + \frac{2L}{nh^2} = O\left(\sqrt{\frac{\log n}{nh^2}}\right).$$

Combining the analysis in Example 5.1, we have

$$\begin{aligned}
\|\widehat{p} - p\|_\infty &\le \underbrace{\Big|\|\widehat{p} - \mathbb{E}[\widehat{p}]\|_\infty - \mathbb{E}\big[\|\widehat{p} - \mathbb{E}[\widehat{p}]\|_\infty\big]\Big|}_{\text{McDiarmid inequality}} + \underbrace{\mathbb{E}\big[\|\widehat{p} - \mathbb{E}[\widehat{p}]\|_\infty\big]}_{\text{Maximal inequality}} \\
&\quad + \underbrace{\|\mathbb{E}[\widehat{p}] - p\|_\infty}_{\text{Bias}} \\
&= O_P\left(\frac{1}{\sqrt{nh^2}}\right) + O\left(\sqrt{\frac{\log n}{nh^2}}\right) + O(h^2) = O_P\left(\sqrt{\frac{\log n}{nh^2}} + h^2\right).
\end{aligned}$$

Choosing the bandwidth $h \sim n^{-1/6}$, we have $\|\widehat{p} - p\|_\infty = O_P(\sqrt{\log n} \cdot n^{-1/3})$.

Part III
High-Dimensional Statistics

Chapter 7
Ordinary Least Squares

7.1 Linear Regression

Given the outcome Y_i and the covariates X_i for $i = 1, \ldots, n$, a regression model assumes

$$Y_i = f(X_i) + \varepsilon_i, \text{ for all } i = 1, \ldots, n,$$

where ε_i is the error/noise. We typically assume that the error terms satisfy $\mathbb{E}\varepsilon_i = 0$ and $\epsilon_1, \ldots, \epsilon_n$ are independent.

A special regression model is the **linear regression model**, where we assume $f(x) = x^\top \beta$ for $\beta \in \mathbb{R}^d$, and thus, the regression model becomes

$$Y_i = X_i^\top \beta + \varepsilon_i, \text{ for all } i = 1, \ldots, n.$$

We also want to introduce some matrix notations. We define the **design matrix** $\mathbb{X} = (X_1^\top, \ldots, X_n^\top)^\top \in \mathbb{R}^{n \times d}$, the **response vector** $Y = (Y_1, \ldots, Y_n)^\top \in \mathbb{R}^n$, and the **noise vector** $\varepsilon = (\varepsilon_1, \ldots, \varepsilon_n)^\top \in \mathbb{R}^n$. We can then write the linear model as

$$Y = \mathbb{X}\beta + \varepsilon.$$

We will also write the design matrix as the columns $\mathbb{X} = (\widetilde{X}_1, \ldots, \widetilde{X}_d)$, where $\widetilde{X}_j$ is the jth column of $\mathbb{X}$.

There are two typical settings for the linear regression:

- **Fixed design**: the features $X_1, \ldots, X_n$ are deterministic.
- **Random design**: the features $X_1, \ldots, X_n$ are random, and we typically assume ε is independent to $\mathbb{X}$.

In this course, we focus on the fixed design setting. Without further specification, we will assume $\mathbb{X}$ is deterministic by default. If $\mathbb{X}$ were actually random, we could

J. Lu, *Big Data Analysis*, https://doi.org/10.1007/978-3-032-03161-7_7

condition the design $\mathbb{X}$ and reduce the problem to a fixed design settings in most situations.

There are two major goals in the study of linear regression:

1. **Prediction**. The regression model should predict outcomes well. We can measure the prediction accuracy by the mean squared error of some estimator $\widehat{f}$ of the true function f^*:

$$\mathrm{MSE}(\widehat{f}) = \frac{1}{n}\sum_{i=1}^{n}\left(\widehat{f}(X_i) - f^*(X_i)\right)^2.$$

   We say an estimator $\widehat{f}$ is **persistent** if $\mathrm{MSE}(\widehat{f}) = o_P(1)$.

   In the linear regression scenario, we can write the mean squared error as

$$\mathrm{MSE}(\mathbb{X}\widehat{\beta}) = \frac{1}{n}\sum_{i=1}^{n}\left(X_i^\top\widehat{\beta} - X_i^\top\beta^*\right)^2 = \frac{1}{n}\|\mathbb{X}(\widehat{\beta}-\beta^*)\|^2 = (\widehat{\beta}-\beta^*)^\top\widehat{\Sigma}(\widehat{\beta}-\beta^*),$$

   where $\widehat{\Sigma} = \mathbb{X}^\top\mathbb{X}/n = \frac{1}{n}\sum_{i=1}^{n} X_i^\top X_i$ is the sample covariance matrix.
2. **Parameter estimation**. Parameter estimates should be consistent. In the linear case, this means looking at the rate of $\|\widehat{\beta} - \beta^*\|$.

7.2 Ordinary Least Squares

We now start with the most popular estimator for the linear model, typically known as the **ordinary least squares** estimator:

$$\widehat{\beta}^{\mathrm{LS}} = \arg\min_{\beta}\sum_{i=1}^{n}(Y_i - X_i^\top\beta)^2 = \arg\min_{\beta}\|Y - \mathbb{X}\beta\|^2.$$

The following proposition gives the closed form formula for the ordinary least squares.

Proposition 7.1 (Closed Form of Least Squares) *The ordinary least squares solution can be explicitly specified by*

$$\widehat{\beta}^{\mathrm{LS}} = (\mathbb{X}^\top\mathbb{X})^\dagger\mathbb{X}^\top Y,$$

where $A^\dagger$ is the Moore-Penrose pseudo-inverse of A.

Proof By definition, the critical points of the least squares loss at $\widehat{\beta}^{\mathrm{LS}}$ have

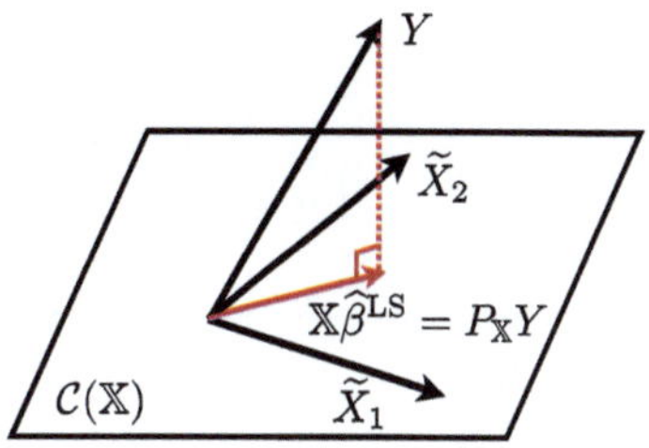

Fig. 7.1 Ordinary least squares is the projection of Y onto $C(\mathbb{X})$

$$0 = \frac{\partial}{\partial \beta} \|Y - \mathbb{X}\beta\|^2 \Big|_{\beta = \widehat{\beta}^{\text{LS}}} = 2\mathbb{X}^\top (Y - \mathbb{X}\widehat{\beta}^{\text{LS}}).$$

Solve the equation above. We have $\mathbb{X}^\top \mathbb{X}\widehat{\beta}^{\text{LS}} = \mathbb{X}^\top Y$ and thus $\widehat{\beta}^{\text{LS}} = (\mathbb{X}^\top \mathbb{X})^\dagger \mathbb{X}^\top Y$. □

Although the closed form of ordinary least squares involves a mysterious pseduo-inverse inside, it actually has a clear geometric interpretation. From Fig. 7.1, the ordinary least squares essentially aim to find a point in the space spanned by $\widetilde{X}_1, \ldots, \widetilde{X}_d$, denoted by $C(\mathbb{X}) := \{\mathbb{X}\beta | \beta \in \mathbb{R}^d\}$, such that it is closest to Y. The formulation of β^{LS} tells us such point is $\mathbb{X}\widehat{\beta}^{\text{LS}}$. Using the geometric language, we say $\mathbb{X}\widehat{\beta}^{\text{LS}}$ is the projection of Y onto the plane $C(\mathbb{X})$, i.e.,

$$\text{Projection } Y \text{ onto } C(\mathbb{X}) = \mathbb{X}\widehat{\beta}^{\text{LS}} = \mathbb{X}(\mathbb{X}^\top \mathbb{X})^\dagger \mathbb{X}^\top Y.$$

Therefore, we define the **projection matrix** $P_{\mathbb{X}} = \mathbb{X}(\mathbb{X}^\top \mathbb{X})^\dagger \mathbb{X}^\top$. It maps any vector to its projection onto the linear space spanned by the columns of $\mathbb{X}$ (see Fig. 7.1). For example, let us consider the projection on a vector $X \in \mathbb{R}^d$. The projection matrix $P_X = XX^\top / \|X\|^2$. So projecting a vector Y onto the vector X can be written as

$$P_X Y = \frac{XX^\top}{\|X\|^2} Y = \frac{\langle X, Y \rangle}{\langle X, X \rangle} X = \Big\langle Y, \frac{X}{\|X\|} \Big\rangle \frac{X}{\|X\|},$$

which is consistent with the geometric concept of a projection.

Not only does the entire vector $\widehat{\beta}^{\text{LS}}$ has a geometric interpretation. In fact, the following proposition gives the geometric meaning of each entry $\widehat{\beta}^{\text{LS}}$.

Proposition 7.2 (Geometry of $\widehat{\beta}_j^{\text{LS}}$) *Let $\mathbb{X}_{-j} = (\widetilde{X}_i, \ldots, \widetilde{X}_{j-1}, \widetilde{X}_{j+1}, \ldots, \widetilde{X}_d) \in \mathbb{R}^{n \times (d-1)}$ denote the design matrix with the jth column deleted. Define*

$$X_j^\perp = X_j - P_{\mathbb{X}_{-j}} X_j,$$

where $X_j^\perp$ is the projection of X_j on the linear space orthogonal to $C(\mathbb{X}_{-j})$. See the vector in Fig. 7.2. We have

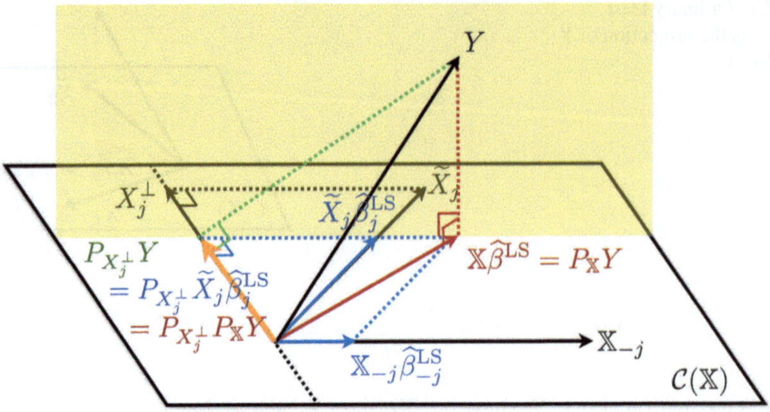

Fig. 7.2 Illustration of Proposition 7.2 and its proof

$$\widehat{\beta}_j^{\mathrm{LS}} = \frac{\left\langle Y, X_j^\perp \right\rangle}{\left\langle X_j^\perp, X_j^\perp \right\rangle}.$$

Namely, $\widehat{\beta}_j^{\mathrm{LS}}$ *is the coefficient assigned to* $X_j^\perp$ *when projecting* Y *onto the vector* $X_j^\perp$*.*

Proof The proof is straightforward in Fig. 7.2. Applying the concept of projection and Euclidean geometry, the key observations in Fig. 7.2 are that the projection of Y onto $X_j^\perp$ (green dashed line) equals to the sequential projections of Y onto $C(\mathbb{X})$ (red dashed line) and then onto $X_j^\perp$ (blue dashed line).

Therefore, see in Fig. 7.2 that the projection of Y (black vector), $P_{\mathbb{X}}Y$ (red vector), and $X_j\widehat{\beta}_j^{\mathrm{LS}}$ (blue vector) onto $X_j^\perp$ are the same orange vector, i.e., the following three vectors are identical:

$$P_{X_j^\perp}Y = P_{X_j^\perp}P_{\mathbb{X}}Y = P_{X_j^\perp}X_j\widehat{\beta}_j^{\mathrm{LS}}.$$

As $P_{X_j^\perp}X_j = X_j^\perp$ by the definition of projection, we have $P_{X_j^\perp}Y = \widehat{\beta}_j^{\mathrm{LS}}X_j^\perp$, i.e.,

$$\frac{\left\langle Y, X_j^\perp \right\rangle}{\left\langle X_j^\perp, X_j^\perp \right\rangle}X_j^\perp = \widehat{\beta}_j^{\mathrm{LS}}X_j^\perp,$$

which completes our proof. □

The last theorem of this chapter aims to show the statistical rate of mean squared error for the ordinary least squares.

Theorem 7.1 (Mean Squared Error of Least Squares) *If the independent noises $\varepsilon_1, \ldots, \varepsilon_n$ have $\mathbb{E}\varepsilon_i = 0$ and ε_i is sub-Gaussian with variance proxy σ^2 for all $i = 1, \ldots, n$ and $\mathrm{rank}(\mathbb{X}) = r$, then*

$$\mathbb{E}\big[\mathrm{MSE}(\mathbb{X}\widehat{\beta}^{\mathrm{LS}})\big] \lesssim \frac{\sigma^2 r}{n},$$

and with probability at least $1 - \delta$,

$$\mathrm{MSE}(\mathbb{X}\widehat{\beta}^{\mathrm{LS}}) \lesssim \frac{\sigma^2 r}{n} + \frac{\sigma^2}{n}\log\left(\frac{1}{\delta}\right).$$

Here we denote $a_n \lesssim b_n$ if there exists a universal constant C irrelevant to n or any other parameters such that $a_n \leq C b_n$ for all n.

Proof The proof strategy is not using the closed form of the least squares. Instead, we directly use the fact that $\widehat{\beta}^{\mathrm{LS}}$ is the minimizer of the sum of squared errors, which is generally called zero-order condition in optimization. This gives us

$$\|Y - \mathbb{X}\widehat{\beta}^{\mathrm{LS}}\|^2 \leq \|Y - \mathbb{X}\beta^*\|^2 = \|\mathbb{X}\beta^* + \varepsilon - \mathbb{X}\beta^*\|^2 = \|\varepsilon\|^2,$$

while on the other side, we have

$$\|Y - \mathbb{X}\widehat{\beta}^{\mathrm{LS}}\|^2 = \|\mathbb{X}\beta^* + \varepsilon - \mathbb{X}\widehat{\beta}^{\mathrm{LS}}\|^2 = \|\mathbb{X}(\widehat{\beta} - \beta^*)\|^2 - 2\big\langle\varepsilon, \mathbb{X}(\widehat{\beta} - \beta^*)\big\rangle + \|\varepsilon\|^2.$$

Consequently, combining the two inequalities above, we have

$$\|\mathbb{X}(\widehat{\beta} - \beta^*)\|^2 \leq 2\langle\varepsilon, \mathbb{X}(\widehat{\beta} - \beta^*)\rangle = 2\|\mathbb{X}(\widehat{\beta} - \beta^*)\|\left\langle\varepsilon, \frac{\mathbb{X}(\widehat{\beta} - \beta^*)}{\|\mathbb{X}(\widehat{\beta} - \beta^*)\|}\right\rangle. \tag{7.1}$$

In the following part of our proof, we will bound the right-hand side of the above inequality via the "sup-out" trick. Let $\Phi = (\phi_1, \ldots, \phi_r) \in \mathbb{R}^{n\times r}$ be an orthogonal matrix whose columns are the orthonormal bases of $C(\mathbb{X})$ such that $\Phi^\top\Phi = \mathrm{I}_r$. Since $\mathbb{X}(\widehat{\beta} - \beta^*) \in C(\mathbb{X})$, there exists $\nu = (\nu_1, \ldots, \nu_r)^\top \in \mathbb{R}^r$ such that $\mathbb{X}(\widehat{\beta} - \beta^*) = \sum_{j=1}^r \nu_j\phi_j = \Phi\nu$.

Therefore, denote $\widetilde{\varepsilon} = \Phi^\top\varepsilon \in \mathbb{R}^r$ and we have

$$\left\langle\varepsilon, \frac{\mathbb{X}(\widehat{\beta} - \beta^*)}{\|\mathbb{X}(\widehat{\beta} - \beta^*)\|}\right\rangle = \left\langle\varepsilon, \frac{\Phi\nu}{\|\Phi\nu\|_2}\right\rangle = \frac{\varepsilon^\top\Phi\nu}{\|\nu\|} = \left\langle\Phi^\top\varepsilon, \frac{\nu}{\|\nu\|}\right\rangle \leq \sup_{\|u\|\leq 1}\langle\widetilde{\varepsilon}, u\rangle = \|\widetilde{\varepsilon}\|,$$

where we use the "sup-out" trick in the first inequality above.

Combining the above inequality with (7.1), we have

$$\mathrm{MSE}(\mathbb{X}\widehat{\beta}^{\mathrm{LS}}) = \frac{1}{n}\|\mathbb{X}(\widehat{\beta} - \beta^*)\|^2 = \frac{4}{n}\left\langle\varepsilon, \frac{\mathbb{X}(\widehat{\beta} - \beta^*)}{\|\mathbb{X}(\widehat{\beta} - \beta^*)\|}\right\rangle^2 \leq \frac{4\|\widetilde{\varepsilon}\|^2}{n}. \tag{7.2}$$

Therefore, we can control the expectation of MSE by

$$\mathbb{E}[\mathrm{MSE}(\mathbb{X}\widehat{\beta}^{\mathrm{LS}})] \le \frac{4\mathbb{E}\|\widetilde{\varepsilon}\|^2}{n} = \frac{4}{n}\sum_{i=1}^{r}\mathbb{E}[\widetilde{\varepsilon}_i^2] \le \frac{4\sigma^2 r}{n},$$

where we use the fact that $\mathbb{E}[\widetilde{\varepsilon}_i^2] = \mathbb{E}(\phi_i^\top \varepsilon)^2 \le \sigma^2$ as the variance is smaller than the variance proxy for sub-Gaussian.

To prove the tail probability inequality of MSE, we need to apply Theorem 6.3, i.e., the maximal inequality for ℓ_2-norm. To implement that theorem, we need to verify that, for any $u \in \mathbb{R}^r$, $\langle u, \widetilde{\varepsilon}\rangle$ is sub-Gaussian with variance proxy $\sigma^2\|u\|^2$, which is true as

$$\mathbb{E}e^{\lambda\langle u,\widetilde{\varepsilon}\rangle} = \mathbb{E}e^{\lambda\langle u,\Phi^\top\varepsilon\rangle} = \mathbb{E}e^{\lambda\langle \Phi u,\varepsilon\rangle} \le e^{\frac{\lambda^2}{2}\|\Phi u\|^2\sigma^2} = e^{\frac{\lambda^2}{2}\sigma^2\|u\|^2},$$

where in the first inequality, we use the fact that $\varepsilon_1, \ldots, \varepsilon_n$ are independent, and we use $\Phi^\top\Phi = \mathrm{I}_r$ in the last inequality.

By (7.2), we have with probability at least $1-\delta$

$$\mathrm{MSE}(\mathbb{X}\widehat{\beta}^{\mathrm{LS}}) \le \frac{4\|\widetilde{\varepsilon}\|^2}{n} \le \frac{4}{n}\left[4\sigma\sqrt{r} + 2\sigma\sqrt{2\log(1/\delta)}\right]^2 \lesssim \frac{\sigma^2 r}{n} + \frac{\sigma^2}{n}\log\left(\frac{1}{\delta}\right),$$

where we use Theorem 6.3 in the second inequality above. □

Chapter 8
Compressive Sensing

8.1 High-Dimensional Linear Models

In the high-dimensional setting, we're essentially looking at the same linear model $Y = \mathbb{X}\beta + \varepsilon$ with $\mathbb{X} \in \mathbb{R}^{n\times d}$. However, we now expect the number of features d is much larger than its sample size n. Under the high-dimensional setting, the ordinary least squares estimator will have troubles. If the features are linearly independent, we have $\text{rank}(\mathbb{X}) = n$. Then, $\mathbb{X}\widehat{\beta}^{\text{LS}} = P_{\mathbb{X}}Y = Y$, i.e., the ordinary least squares will overfit. Therefore, we need to invoke the parsimonious principle and introduce the following sparse linear model.

Definition 8.1 (Sparse Linear Model) A sparse linear model satisfies $Y = \mathbb{X}\beta + \varepsilon$ with $\mathbb{X} \in \mathbb{R}^{n\times d}$, but only s entries of β are nonzero. We also define the ℓ_0-norm

$$\|\beta\|_0 = \sum_{j=1}^{d} \mathbb{1}\{\beta_j \neq 0\} = s.$$

Typically, the number of effective features s is much smaller than d. So the estimation of sparse linear model is like finding a needle in a haystack. A straightforward method to estimate the sparse linear model is to modify the least squares by adding the sparsity constraint:

$$\min_{\beta} \|Y - \mathbb{X}\beta\|_2^2, \text{ s.t. } \|\beta\|_0 \leq s.$$

However, the sparsity constraint $\|\beta\|_0 \leq s$ is very complicated and will cause big troubles in computation. In fact, we can decompose the constraint as

$$\left\{\beta \in \mathbb{R}^d \,|\, \|\beta\|_0 \leq s\right\} = \bigcup_{\mathcal{S}:|\mathcal{S}|=s} \left\{\beta \in \mathbb{R}^d | \beta_j \neq 0 \text{ for all } j \in \mathcal{S} \subseteq \{1, \ldots, d\}\right\}.$$

J. Lu, *Big Data Analysis*, https://doi.org/10.1007/978-3-032-03161-7_8

From the decomposition above, there are $\binom{d}{s}$ possible subspace $\mathcal{S}$, which makes us have to solve $\binom{d}{s}$ least squares. This is not acceptable when d is very large.

An alternative idea is to relax the computationally infeasible ℓ_0-norm to some other norms which are computationally friendly. This motivates us to introduce the following ℓ_p-norms.

Definition 8.2 (ℓ_p-Norm) The ℓ_p-norm of a d-dimensional vector $x = (x_1, \ldots, x_d)^\top$ is defined as

$$\|\boldsymbol{x}\|_p = \Big(\sum_{j=1}^{d} |x_j|^p\Big)^{1/p}, \text{ for } p \in (0, \infty).$$

We also define the ℓ_∞-norm as $\|x\|_\infty = \max_{1\le j\le d} |x_j|$.

We can see that the ℓ_2-norm used in the previous chapters are consistent with the definition above. Therefore, in the following parts of our book, we will add the subscript p to $\|\cdot\|$ to emphasize which ℓ_p-norm we are using. Notice that the triangular inequality $\|x+y\|_p \le \|x\|_p + \|y\|_p$ is only true when $p \ge 1$. Therefore, the ℓ_p-norm for $p \ge 1$ is a real norm, while ℓ_p-norm for $p < 1$ is only a pseudo-norm. Similar to the variational form of the ℓ_2-norm: $\|x\|_2 = \sup_{\|y\|_2\le 1}\langle x, y\rangle$, we also have the following variational form of general ℓ_p-norms for $p \ge 1$.

Proposition 8.1 (Variational Form of ℓ_p-Norm) *For any* $1 \le p \le \infty$, *we have*

$$\|x\|_p = \sup_{\|y\|_q\le 1} \langle x, y\rangle, \textit{ where } \frac{1}{p} + \frac{1}{q} = 1.$$

We call ℓ_q-norm as the dual norm of ℓ_p-norm, and vice versa. For example, ℓ_2-norm is the dual norm of itself. ℓ_∞-norm is the dual norm of ℓ_1-norm, and ℓ_1-norm is the dual norm of ℓ_∞-norm. Proposition 8.1 also implies the following inequality:

$$\text{Hölder inequality: } |\langle x, y\rangle| \le \|x\|_p\|y\|_q \text{ for any } \frac{1}{p} + \frac{1}{q} = 1.$$

We can see that the Cauchy-Schwarz inequality $|\langle x, y\rangle| \le \|x\|_2\|y\|_2$ is a special case of the Hölder inequality for $p = 2$.

Now let us study the geometry of ℓ_p-norms. We visualize the ℓ_p-ball $\{x \mid \|x\|_p \le 1\}$ in Fig. 8.1. In Fig. 8.1a, we can see that the ℓ_1-ball has a diamond shape and the ℓ_∞-ball is a square. When p increases from 1 to ∞, the diamond gradually expands to a square. In Fig. 8.1b, we can see that when $p < 1$, the ℓ_p-ball has an asteroid shape, and it keeps shrinking as p goes to zero.

From Fig. 8.1, we can see that the ℓ_1-ball is the smallest convex ball. Therefore, the ℓ_1-norm is the *convex relaxation* of ℓ_1-norm. This leads us to estimating the high-dimensional linear model via the following convex problem:

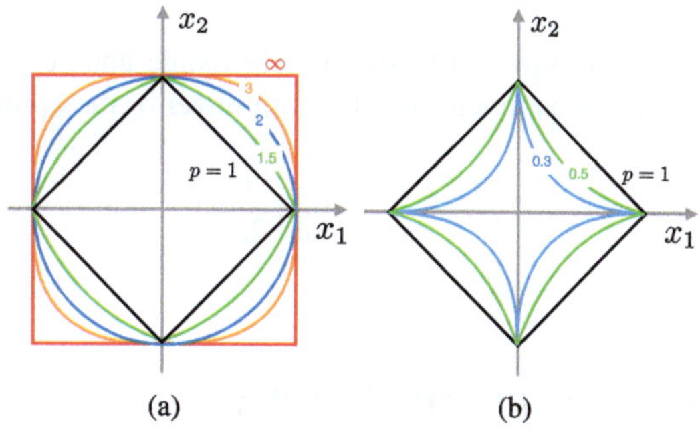

Fig. 8.1 The ℓ_p-ball $\{x | \, \|x\|_p \leq 1\}$ in two-dimensional space. (**a**) $p \geq 1$. (**b**) $p \leq 1$

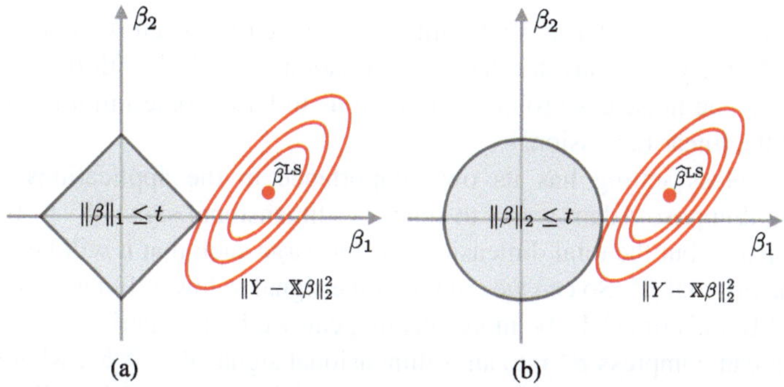

Fig. 8.2 The geometry of the optimization problem in (8.1). (**a**) ℓ_1 constraint. (**b**) ℓ_2 constraint

$$\min_{\beta} \|Y - \mathbb{X}\beta\|_2^2, \text{ s.t. } \|\beta\|_1 \leq t, \tag{8.1}$$

where t is some tuning parameter. We visualize the above problem in Fig. 8.2a. Compared to the ℓ_2-norm in Fig. 8.2b, the diamond shape of ℓ_1-ball tends to make the solution of the optimization problem located on the vertices of the polytope, which correspond to sparse vectors. This geometric insight also explains why the ℓ_1-norm seems not directly encouraging sparsity like ℓ_0-norm by its definition, but it can actually help us to the sparse solutions. On the other hand, the optimization problem on the convex ℓ_1-norm is much more feasible than the ℓ_0-norm. That is why ℓ_1-norm is so welcomed in the high-dimensional statistics: it can help us to select variables, and it is easy in computation.

Applying the method of Lagrange multipliers, we know that for any $t > 0$, there exists a λ such that the constraint optimization problem in (8.1) is equivalent to the following unconstrained optimization problem.

The **Lasso** (least absolute shrinkage and selection operator) estimator with tuning parameter λ is the minimizer of the optimization problem

$$\min_{\beta} \frac{1}{2n}\|Y - \mathbb{X}\beta\|_2^2 + \lambda\|\beta\|_1.$$

8.2 Compressive Sensing

Before studying Lasso, let us start with a simpler setting when the linear model is noiseless. Namely, we consider the linear equation $Y = \mathbb{X}\beta^*$ with the sparse truth $\|\beta^*\|_0 \leq s$. The noiseless problem can help us to build some intuition on the ℓ_1-norm and the linear regression.

The noiseless setting has its own importance in the applications of signal processing. Suppose Adam wants to send a d-dimensional signal β^* to Bella. The signal is sparse, but the total dimension d is too large such that it will be inefficient to send the original β^*. So can we compress the signal to lower dimensions and save Adam and Bella's time? To be more specific, can we find a matrix $\mathbb{X} \in \mathbb{R}^{n\times d}$ such that Adam can compress β^* into an n-dimensional signal $Y = \mathbb{X}\beta^*$, where n could be much smaller than d. Instead of sending β^*, Adam can send Y to Bella, which is more efficient. If Bella has the matrix $\mathbb{X}$ priorly, she can recover the β^* via some compressive sensing algorithm. We then have two major problems in compressive sensing:

1. What is the algorithm to recover β^*?
2. What kind of matrix $\mathbb{X}$ can guarantee the recovery?
3. How efficiently can we compress β^*, i.e., how small can n be with respect to d?

Let us start with the first question. Ideally, we can find the true β^* by finding the sparsest β satisfying the linear equation $Y = \mathbb{X}\beta$, i.e., we consider the optimization problem

$$\min_{\beta} \|\beta\|_0, \text{ s.t. } Y = \mathbb{X}\beta.$$

However, we know that this problem is challenging in computation. We know that all possible solutions of the linear equation $Y = \mathbb{X}\beta$ are

$$\beta = (\mathbb{X}^\top\mathbb{X})^\dagger\mathbb{X}Y + \Delta, \text{ for any } \Delta \in \text{Null}(\mathbb{X}),$$

where $\text{Null}(\mathbb{X}) = \{\Delta |\ \mathbb{X}\Delta = 0\}$. However, it is challenging to find the sparsest vector among these solutions as $(\mathbb{X}^\top\mathbb{X})^\dagger\mathbb{X}Y$ may not be sparse.

Similar to the discussion in the previous section, we can relax the ℓ_0-norm to ℓ_1-norm and consider the following estimator.

The **basis pursuit** estimator is the minimizer of the linear optimization problem

$$\widehat{\beta} = \arg\min_{\beta} \|\beta\|_1, \text{ s.t. } Y = \mathbb{X}\beta.$$

We can solve this problem efficiently via linear programming algorithms.

Now we study when the basis pursuit can recover the truth, i.e., $\widehat{\beta} = \beta^*$. From Fig. 8.3a, we can see that if the linear space $\{\beta | Y = \mathbb{X}\beta\}$ is tangent to the ℓ_1-ball $\{\beta | \; \|\beta\|_1 \leq \|\beta^*\|_1\}$, the basis pursuit estimator $\widehat{\beta}$ locates at the tangent point β^*, i.e., we have a perfect recovery. On the other hand, in Fig. 8.3b, if the linear space $\{\beta | Y = \mathbb{X}\beta\}$ gets into the ℓ_1-ball, $\widehat{\beta}$ does not equal to β^*. Therefore, Fig. 8.3 shows us that the basis pursuit can recover the truth if and only if the null space Null($\mathbb{X}$) does not locate in the yellow area. In fact, the yellow area is a cone defined as follows.

Definition 8.3 (Cone Condition) We denote the support of β^* as $S = \{j | \beta_j^* \neq 0\}$. We also denote $\beta_S \in \mathbb{R}^{|S|}$ as a sub-vector of β, whose entries are β_j's for all $j \in S$. Define the cone

$$\mathbb{C}(S) = \{\Delta | \; \|\Delta_{S^c}\|_1 \leq \|\Delta_S\|_1\}.$$

The cone $\mathbb{C}(S)$ corresponds to the yellow area in Fig. 8.3. It contains the vectors whose ℓ_1-norm not on the support is dominated by the ℓ_1-norm on the support.

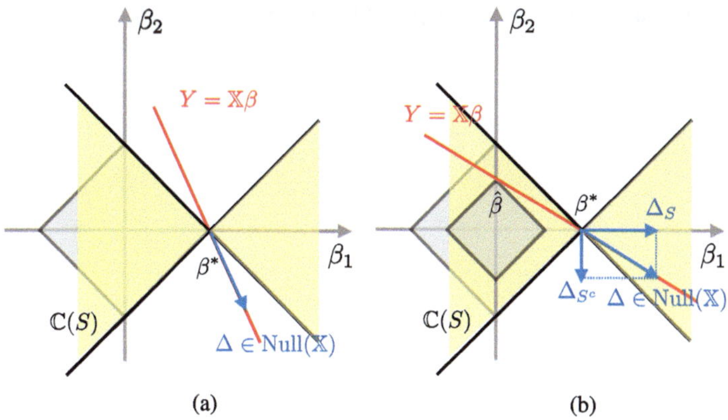

Fig. 8.3 The geometry of the basis pursuit and the cone condition. (**a**) Perfect recovery. (**b**) No perfect recovery

The following theorem formalizes our intuition in Fig. 8.3 and gives a sufficient and necessary condition of when the basis pursuit is perfect recovery.

Theorem 8.1 *The basis pursuit has a unique solution* $\widehat{\beta} = \beta^*$ *for any* $\beta^* \in \mathbb{R}^d$ *with support* S *if and only if*

$$\mathbb{C}(S) \bigcap \text{Null}(\mathbb{X}) = \{0\}. \tag{8.2}$$

Proof We first prove that (8.2) is a sufficient condition for perfect recovery.

Let $\widehat{\Delta} = \widehat{\beta} - \beta^*$. Therefore, $\widehat{\Delta} \in \text{Null}(\mathbb{X})$ as $\mathbb{X}(\widehat{\beta} - \beta^*) = 0$. We aim to show that $\widehat{\Delta} \in \mathbb{C}(S)$ and by (8.2), $\widehat{\Delta} = 0$. By the definition of basis pursuit, we have $\|\beta^*\|_1 \geq \|\widehat{\beta}\|_1$. Moreover, we have

$$\begin{aligned} \|\beta_S^*\|_1 = \|\beta^*\|_1 \geq \|\widehat{\beta}\|_1 &= \|\beta^* + \widehat{\Delta}\|_1 \\ &= \|\beta_S^* + \widehat{\Delta}_S\|_1 + \|\widehat{\Delta}_{S^c}\|_1 \\ &\geq \|\beta_S^*\|_1 - \|\widehat{\Delta}_S\|_1 + \|\widehat{\Delta}_{S^c}\|_1. \end{aligned} \tag{8.3}$$

This implies that $\|\widehat{\Delta}_S\|_1 \geq \|\widehat{\Delta}_{S^c}\|_1$. Hence, $\widehat{\Delta} \in \mathbb{C}(S)$ and by (8.2), $\widehat{\beta} = \beta^*$.

Now we prove that (8.2) is necessary for the perfect recovery. We aim to show that if $\widehat{\beta} = \beta^*$ for all β^* such that $\|\beta^*\|_0 \leq s$, then $\mathbb{C}(S) \bigcap \text{Null}(\mathbb{X}) = \{0\}$.

For any $\beta^* \in \text{Null}(\mathbb{X}) \backslash \{0\}$, our goal is to show that $\beta^* \notin \mathbb{C}(S)$. As the basis pursuit can always perfectly recover the truth, we have

$$\begin{pmatrix} \beta_S^* \\ 0 \end{pmatrix} = \mathop{\arg\min}_{\beta \in \mathbb{R}^d} \|\beta\|_1, \text{ s.t. } \mathbb{X}\beta = \mathbb{X} \begin{pmatrix} \beta_S^* \\ 0 \end{pmatrix}.$$

As $\mathbb{X}\beta^* = 0$, the vector $(0, -\beta_{S^c}^\top)^\top$ is also a feasible solution to the problem. Therefore, by uniqueness of basis pursuit, we have $\|\beta_S^*\|_1 < \|\beta_{S^c}^*\|_1$, i.e., $\beta^* \notin \mathbb{C}(S)$. □

Chapter 9
Restricted Isometry Property

9.1 Restricted Isometry Property

In the previous chapter, we introduce the problem of compressive sensing: how to find the sparse truth β^* from the linear equation $Y = \mathbb{X}\beta^*$. Recall that we list three major questions for the compressive sensing:

1. What is the algorithm to recover β^*?
2. What kind of matrix $\mathbb{X}$ can guarantee the recovery?
3. How efficiently can we compress β^*, i.e., how small can n be with respect to d?

The first question is solved by the basis pursuit estimator $\hat{\beta} = \arg\min_\beta \|\beta\|_1$ such that $Y = \mathbb{X}\beta$. The second question is partially answered in Theorem 8.1, as we show that the cone condition $\mathbb{C}(S) \cap \mathrm{Null}(\mathbb{X}) = 0$ is a sufficient and necessary condition for the perfect recovery of basis pursuit in Theorem 8.1. However, the cone condition is not easy to use in practice. It is not straightforward to construct $\mathbb{X}$ starting from the cone condition. In this chapter, we will discuss another sufficient condition for perfect recovery, called restricted isometry property, which is stronger but easier to implement. We will talk about how to construct $\mathbb{X}$ based on this property and answer the third question.

Definition 9.1 (Restricted Isometry Property) We say $\mathbb{X}$ satisfies s-RIP with the coefficient $\delta_s \in (0, 1)$ if, for any β with $\|\beta\|_0 \le s$, we have

$$(1 - \delta_s)\|\beta\|_2^2 \le \|\mathbb{X}\beta\|_2^2 \le (1 + \delta_s)\|\beta\|_2^2.$$

The RIP condition implies that the linear map $\mathbb{X}$ only distorts the norm of s-sparse vectors by δ_s, i.e.,

$$\left|\frac{\|\mathbb{X}\beta\|_2^2}{\|\beta\|_2^2} - 1\right| = \left|\frac{\beta^\top \mathbb{X}^\top \mathbb{X}\beta}{\|\beta\|_2^2} - 1\right| \le \delta_s, \text{ for all } \|\beta\|_0 \le s.$$

J. Lu, *Big Data Analysis*, https://doi.org/10.1007/978-3-032-03161-7_9

Since $\mathbb{X}^\top Y = \mathbb{X}^\top\mathbb{X}\beta^*$, ideally, if $\mathbb{X}^\top\mathbb{X} = \mathrm{I}_d$, then we can directly recover the β^* by $\mathbb{X}^\top Y$. However, in high-dimensional setting, we have $\mathrm{rank}(\mathbb{X}^\top\mathbb{X}) \le n \ll d$; thus, it is impossible for $\mathbb{X}^\top\mathbb{X}$ to be an identity matrix. However, we know β^* is s-sparse, so we only need $\mathbb{X}^\top\mathbb{X}$ to be somehow identity for s-sparse vectors. In that sense, the s-RIP condition assumes that $\mathbb{X}^\top\mathbb{X}$ is an "almost" identity matrix for all s-sparse vectors. And the level of almostness is measured by δ_s.

Therefore, the discussion above provides us the intuition that we can recover the truth if $\mathbb{X}^\top\mathbb{X}$ is almost an identity. We now formalize this intuition by the theorem below.

Theorem 9.1 (Perfect Recovery Under RIP) *If $\mathbb{X}$ is 3s-RIP with the coefficient $\delta_s < \frac{1}{3}$, then the basis pursuit can perfectly recover the truth, i.e., $\widehat{\beta} = \beta^*$.*

Proof Let $\Delta = \widehat{\beta} - \beta^*$. Therefore, $\mathbb{X}(\beta - \beta^*) = 0$ and $\Delta \in \mathrm{Null}(\mathbb{X})$. Same as Eq. (8.3) in the proof of Theorem 8.1 of Chap. 8 , we have

$$\begin{aligned}\|\beta_S^*\|_1 = \|\beta^*\|_1 \ge \|\widehat{\beta}\|_1 &= \|\beta^* + \Delta\|_1\\ &= \|\beta_S^* + \Delta_S\|_1 + \|\Delta_{S^c}\|_1\\ &\ge \|\beta_S^*\|_1 - \|\Delta_S\|_1 + \|\Delta_{S^c}\|_1.\end{aligned}$$

Thus, $\|\widehat{\Delta}_{S^c}\|_1 \le \|\widehat{\Delta}_S\|_1$. If we can also show that $\|\Delta_S\|_1 \le \rho\|\Delta_{S^c}\|_1$ for some $\rho < 1$, then Δ must be zero and $\widehat{\beta} = \beta^*$. Therefore, to complete the proof, it suffices to find a $\rho < 1$ and prove the inequality $\|\Delta_S\|_1 \le \rho\|\Delta_{S^c}\|_1$ under $3s$-RIP.

Our idea is to apply the $3s$-RIP to Δ. However, the RIP condition only applies to $3s$-sparse vectors. Therefore, our first step is to divide Δ into shorter sub-vectors as follows:

$$\begin{aligned}S_0 &= S,\\ S_1 &= \text{entries of the } 2s \text{ largest absolute values in } \Delta_{S^c},\\ S_2 &= \text{entries of the next } 2s \text{ largest absolute values in } \Delta_{S^c},\\ &\vdots\end{aligned}$$

Recall that $\Delta \in \mathrm{Null}(\mathbb{X})$ and we have

$$0 = \|\mathbb{X}\Delta\|_2 = \|\mathbb{X}\textstyle\sum_{j\ge0}\Delta_{S_j}\|_2 \ge \|\mathbb{X}(\Delta_{S_0} + \Delta_{S_1})\|_2 - \|\sum_{j\ge2}\mathbb{X}\Delta_{S_j}\|_2.$$

Therefore, $\|\mathbb{X}(\Delta_{S_0} + \Delta_{S_1})\|_2 \le \|\sum_{j\ge2}\mathbb{X}\Delta_{S_j}\|_2$. Applying $3s$-RIP on both sides of this inequality, we have

$$\begin{aligned}\sqrt{(1-\delta_{3s})}\|\Delta_{S_0} + \Delta_{S_1}\|_2 &\le \|\mathbb{X}(\Delta_{S_0} + \Delta_{S_1})\|_2\\ &\le \|\textstyle\sum_{j\ge2}\mathbb{X}\Delta_{S_j}\|_2 \le \sqrt{(1+\delta_{3s})}\sum_{j\ge2}\|\Delta_{S_j}\|_2.\end{aligned}$$

The above result has established an inequality between Δ_S and Δ_{S^c}, but in ℓ_2-norm. Recall that our goal is to prove $\|\Delta_S\|_1 \leq \rho\|\Delta_{S^c}\|_1$, which is in ℓ_1-norm. Therefore, the remaining part of the proof is to bridge between ℓ_2-norm and ℓ_1-norm. We have

$$\begin{aligned}\sqrt{(1-\delta_{3s})}\|\Delta_S+\Delta_{S_1}\|_2 &\leq \sqrt{(1+\delta_{3s})}\sum_{j\geq 2}\|\Delta_{S_j}\|_2\\ &\leq \sqrt{(1+\delta_{3s})}\sum_{j\geq 2}\sqrt{2s}\|\Delta_{S_j}\|_\infty\\ &\leq \sqrt{(1+\delta_{3s})}\sum_{j\geq 2}\frac{1}{\sqrt{2s}}\|\Delta_{S_{j-1}}\|_1=\sqrt{\frac{(1+\delta_{3s})}{2s}}\|\Delta_{S^c}\|_1,\end{aligned}$$

where in the second inequality, we use the fact that $\|x\|_2 \leq \sqrt{d}\|x\|_\infty$, for any $x \in \mathbb{R}^d$, and in the last inequality, we use the inequality $\|\Delta_{S_j}\|_\infty \leq \frac{1}{2s}\|\Delta_{S_{j-1}}\|_1$ for $j \geq 2$ as we construct Δ_{S_j} in order; thus, the maximum norm of Δ_{S_j} is smaller than the average of the preceded sub-vector $\Delta_{S_{j-1}}$.

On the left-hand side of the above inequality, we have

$$\sqrt{(1-\delta_{3s})}\|\Delta_S+\Delta_{S_1}\|_2 \geq \sqrt{(1-\delta_{3s})}\|\Delta_S\|_2 \geq \sqrt{\frac{(1-\delta_{3s})}{s}}\|\Delta_S\|_1,$$

where in the last inequality, we use the fact that $\|x\|_1 \leq \sqrt{d}\|x\|_2$, for any $x \in \mathbb{R}^d$. Summarizing the inequalities above, we have

$$\|\Delta_S\|_1 \leq \sqrt{\frac{(1+\delta_{3s})}{2(1-\delta_{3s})}}\|\Delta_{S^c}\|_1,$$

and therefore, if $\delta_{3s} < 1/3$, we choose $\rho := \sqrt{\frac{(1+\delta_{3s})}{2(1-\delta_{3s})}} < 1$ and $\|\Delta_S\|_1 \leq \rho\|\Delta_{S^c}\|_1$. □

Now we are ready to answer the second and third questions listed in the beginning of this chapter. By Theorem 9.1, it suffices to find $\mathbb{X}$ satisfying $3s$-RIP. Although the $3s$-RIP is a deterministic property of a matrix, the easiest way to find such matrix is to select it randomly. As we discussed above, RIP condition implies that $\mathbb{X}^\top\mathbb{X}$ is "almost" identity. If we can randomly select $\mathbb{X}$ such that $\mathbb{E}[\mathbb{X}^\top\mathbb{X}] = \mathrm{I}_d$, then by the concentration principle, we may expect $\mathbb{X}^\top\mathbb{X}$ has similar property as the identity. The following theorem gives us a concrete method to construct $\mathbb{X}$ satisfying $3s$-RIP.

Theorem 9.2 *Suppose the entries of* $\mathbb{X} \in \mathbb{R}^{n\times d}$ *are i.i.d.* $N\left(0, \frac{1}{n}\right)$. *For any* $\varepsilon > 0$, $\delta \in (0, 1)$, *if* $n \geq \frac{96}{\delta^2}s\log(\frac{18d}{\varepsilon})$, *then*

$$\mathbb{P}\big(\mathbb{X} \textit{ is 3s-RIP with the coefficient } \delta\big) \geq 1-\varepsilon.$$

This theorem answers the questions of how to construct $\mathbb{X}$ and how efficiently we can compress the signal. We can compress a d-dimensional s-sparse signal into $n = O(s \log d)$ dimensions and perfectly recover it via basis pursuit by Theorem 9.1.

Proof By the definition of 3s-RIP, we need to control the probability of the event

$$\left\{\left|\frac{\|\mathbb{X}\beta\|_2^2}{\|\beta\|_2^2} - 1\right| > \delta \text{ for all } \|\beta\|_0 \le 3s\right\} = \left\{\sup_{\|\beta\|_0 \le 3s}\left|\frac{\|\mathbb{X}\beta\|_2^2}{\|\beta\|_2^2} - 1\right| > \delta\right\}.$$

This is an event involving maximums, and we will apply the maximal inequality and the discretization trick. Let us start with a fixed β. Given a $\beta \in \mathbb{R}^d$, we have $X_i^\top \beta$ for $i = 1, \ldots, n$ are i.i.d. $N(0, \frac{1}{n}\|\beta\|_2^2)$. Therefore, we have

$$\frac{\|\mathbb{X}\beta\|_2^2}{\|\beta\|_2^2} = \sum_{i=1}^n \left(\frac{X_i^\top \beta}{\|\beta\|_2^2}\right)^2 \sim \frac{1}{n}\chi_n^2.$$

By the concentration inequality of χ_n^2 (see, e.g., Homework 1, Q2(b)), we have

$$\mathbb{P}\left(\left|\frac{\|\mathbb{X}\beta\|_2^2}{\|\beta\|_2^2} - 1\right| > \delta\right) = \mathbb{P}\left(\left|\frac{\|\mathbb{X}\beta\|_2^2}{\|\beta\|_2^2} - \mathbb{E}\frac{\|\mathbb{X}\beta\|_2^2}{\|\beta\|_2^2}\right| > \delta\right) \le 2e^{-\frac{n\delta^2}{8}}. \tag{9.1}$$

The second step is to apply the ϵ-net. We can decompose the ℓ_0-ball $\|\beta\|_0 \le 3s$ into $\binom{d}{3s}$ number of 3s-dimensional subspaces. Therefore, by the union bound, we have

$$\begin{aligned}\mathbb{P}\left(\sup_{\|\beta\|_0 \le 3s}\left|\frac{\|\mathbb{X}\beta\|_2^2}{\|\beta\|_2^2} - 1\right| > \delta\right) &\le \binom{d}{3s}\mathbb{P}\left(\sup_{\beta \in \mathcal{B}_2^{3s}}\left|\frac{\|\mathbb{X}\beta\|_2^2}{\|\beta\|_2^2} - 1\right| > \delta\right)\\ &= \binom{d}{3s}\mathbb{P}\left(\sup_{u \in \mathcal{B}_2^{3s}}\left|\|\mathbb{X}u\|_2^2 - 1\right| > \delta\right),\end{aligned}$$

where we abuse the notation in the first inequality: when $\beta \in \mathcal{B}_2^{3s}$, the number of columns of $\mathbb{X}$ becomes 3s automatically.

We denote $A = \mathbb{X}^\top\mathbb{X} - \mathrm{I}$ and $\|\mathbb{X}u\|_2^2 - 1 = u^\top(\mathbb{X}^\top\mathbb{X} - \mathrm{I})u = u^\top A u$. Let $\mathcal{N}_{1/4} \subseteq \mathcal{B}_2^{3s}$ be the $\frac{1}{4}$-net of $\mathcal{B}_2^{3s}$. Then for any $u \in \mathcal{B}_2^{3s}$, there exists a $v \in \mathcal{N}_\epsilon$ such that $\|u - v\|_2 \le \frac{1}{4}$. Therefore, we have

$$\begin{aligned}\sup_{u \in \mathcal{B}_2^{3s}} |u^\top A u| &\le \sup_{u \in \mathcal{B}_2^{3s}} |(u - v)^\top A u + v^\top A(u - v) + v^\top A v|\\ &\le \sup_{u \in \mathcal{B}_2^{3s}} \frac{1}{4}|u^\top A u| + \sup_{u \in \mathcal{B}_2^{3s}} \frac{1}{4}|u^\top A u| + \sup_{v \in \mathcal{N}_{1/4}} |v^\top A v|.\end{aligned}$$

Therefore, $\sup_{u \in \mathcal{B}_2^{3s}} |u^\top A u| \le 2 \sup_{v \in \mathcal{N}_{1/4}} |v^\top A v|$ and we have

$$
\begin{aligned}
&\mathbb{P}\left(\sup_{\|\beta\|_0 \le 3s} \left| \frac{\|\mathbb{X}\beta\|_2^2}{\|\beta\|_2^2} - 1 \right| > \delta \right) \\
&\le \binom{d}{3s} \mathbb{P}\left(\sup_{u \in \mathcal{B}_2^{3s}} \left| \|\mathbb{X}u\|_2^2 - 1 \right| > \delta \right) \\
&\le \binom{d}{3s} \mathbb{P}\left(\sup_{v \in \mathcal{N}_{1/4}} \left| \|\mathbb{X}v\|_2^2 - 1 \right| > \frac{\delta}{2} \right) \\
&\le d^{3s} |\mathcal{N}_{1/4}| \cdot \mathbb{P}\left(\left| \frac{\|\mathbb{X}v\|_2^2}{\|v\|_2^2} - 1 \right| > \frac{\delta}{2} \right) \qquad (\text{As } \|v\|_2 = 1) \\
&\le (9d)^{3s} \cdot 2e^{-\frac{n\delta^2}{32}},
\end{aligned}
$$

where in the last inequality, we use the facts (9.1) and $|\mathcal{N}_{1/4}| \le 9^{3s}$ by Lemma 6.1. Solve $\varepsilon \ge 2(9d)^{3s} e^{-\frac{n\delta^2}{32}}$ and we have $n \ge \frac{96}{\delta^2} s \log(\frac{18d}{\varepsilon})$. □

Chapter 10
Statistical Properties of Lasso

10.1 Restricted Eigenvalue Condition

In this chapter, we return to the noisy linear regression. Recall the sparse linear model $Y = \mathbb{X}\beta^* + \varepsilon$, where $\mathbb{X} \in \mathbb{R}^{n\times d}$ and $\|\beta^*\|_0 \leq s$. We estimate the high-dimensional linear model via the Lasso estimator

$$\widehat{\beta} = \arg\min_{\beta} \frac{1}{2n}\|Y - \mathbb{X}\beta\|_2^2 + \lambda\|\beta\|_1.$$

In this chapter, we will study the statistical properties of the Lasso estimator. Like the RIP condition for the basis pursuit, we also need conditions for Lasso.

Definition 10.1 (Restricted Eigenvalue) Let S be the support of β^*. We say $\mathbb{X}$ satisfies the restricted eigenvalue condition RE(κ, α) if

$$\frac{1}{n}\|\mathbb{X}\Delta\|_2^2 \geq \kappa\|\Delta\|_2^2, \text{ for all } \Delta \in \mathbb{C}_\alpha(S) := \{\Delta \mid \|\Delta_{S^c}\|_1 \leq \alpha\|\Delta_S\|_1\}. \tag{10.1}$$

We first compare restricted eigenvalue (RE) condition with the restricted isometry property (RIP). The restricted isometry property states that

$$(1-\delta_s)\|\Delta\|_2^2 \leq \|\mathbb{X}\Delta\|_2^2 \leq (1+\delta_s)\|\Delta\|_2^2,$$

for all $\|\Delta\|_0 \leq s$. Compared to the two-sided inequality on $\|\mathbb{X}\Delta\|_2^2$ in RIP, the restricted eigenvalue condition only impose a one-sided inequality. Moreover, the word "restricted" in RIP means the inequality is truth for $\|\Delta\|_0 \leq s$, but the RE restricts Δ in the cone $\|\Delta_{S^c}\|_1 \leq \alpha\|\Delta_S\|_1$. Therefore, the RE condition is less restricted compared to RIP.

In fact, notice that $\frac{1}{n}\|\mathbb{X}\Delta\|_2^2 = \Delta^\top\widehat{\Sigma}\Delta$, where $\widehat{\Sigma} = \mathbb{X}^\top\mathbb{X}/n$ is the sample covariance matrix of the features. Thus, the RE condition is also related to $\widehat{\Sigma}$. By

J. Lu, *Big Data Analysis*, https://doi.org/10.1007/978-3-032-03161-7_10

Theorem 3.2, the minimum eigenvalue of sample covariance matrix $\lambda_{\min}(\widehat{\Sigma})$ has the following variational form:

$$\lambda_{\min}(\widehat{\Sigma}) = \min_{\Delta} \frac{\Delta^\top \widehat{\Sigma} \Delta}{\|\Delta\|_2^2} = \min_{\Delta} \frac{\Delta^\top \mathbb{X}^\top \mathbb{X} \Delta}{n\|\Delta\|_2^2} = \min_{\Delta} \frac{1}{n} \frac{\|\mathbb{X}\Delta\|_2^2}{\|\Delta\|_2^2}. \tag{10.2}$$

In comparison, the RE condition in (10.1) is equivalent to the formulation

$$\min_{\Delta \in \mathbb{C}_\alpha(S)} \frac{1}{n} \frac{\|\mathbb{X}\Delta\|_2^2}{\|\Delta\|_2^2} \geq \kappa.$$

This is why we call it the restricted eigenvalue condition, as the κ is somehow the minimum eigenvalue of $\widehat{\Sigma}$ restricted on the cone $\mathbb{C}_\alpha(S)$. We know that a matrix A is positive definite, denoted as $A \succ 0$, if $\lambda_{\min}(A) > 0$. However, under the high dimensional setting when $d \gg n$, we have $\text{rank}(\mathbb{X}) \leq n \ll d$ so $\lambda_{\min}(\widehat{\Sigma}) = 0$. Namely, $\widehat{\Sigma}$ is positive semi-definite, denoted as $\widehat{\Sigma} \succeq 0$. In that sense, the RE condition assumes that $\mathbb{X}^\top\mathbb{X}/n$ is "almost" positive definite and the word "almost" means we restricted on the cone. Recall that in the previous chapter, we say RIP assumes that $\mathbb{X}^\top\mathbb{X}$ is almost identity, which is more stringent than positive definiteness. This is because we need perfect recovery in the compressive sensing, while we only expect good statistical rate for Lasso.

Now we provide some intuition why the analysis of Lasso requires the RE condition. This is due to the landscape of the least squares loss $\mathcal{L}_n(\beta) = \frac{1}{2n}\|Y - \mathbb{X}\beta\|_2^2$. By concentration principle, we may expect that $\mathcal{L}_n(\widehat{\beta})$ is close to $\mathcal{L}_n(\beta^*)$. From Fig. 10.1a, we can see that if $\mathcal{L}_n(\beta)$ is convex enough, then $\widehat{\beta}$ is also close to β^*. However, if $\mathcal{L}_n(\beta)$ is very flat, then $\widehat{\beta}$ could be far away from β^* even if $\mathcal{L}_n(\widehat{\beta})$ is close to $\mathcal{L}_n(\beta^*)$. This means that the curvature of $\mathcal{L}_n(\beta)$ around β^* is relatively small. We know that the curvature of a curve can be characterized by the Hessian matrix $\nabla^2\mathcal{L}_n(\widehat{\beta})$. The larger the minimum eigenvalue of the Hessian $\lambda_{\min}(\nabla^2\mathcal{L}_n(\widehat{\beta}))$ is, the more convex $\mathcal{L}_n(\widehat{\beta})$ is. However, for the least squares loss

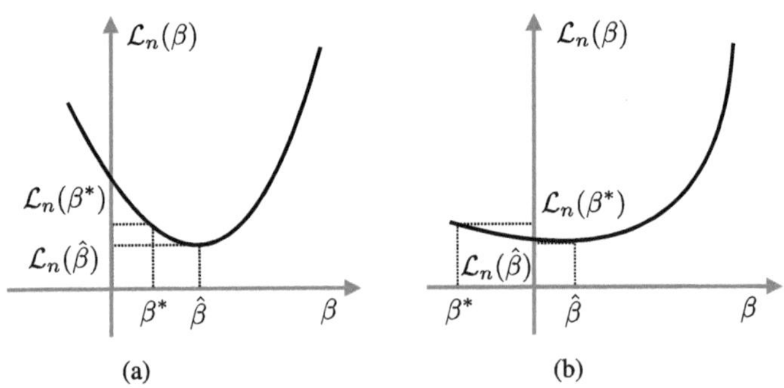

Fig. 10.1 The curvature of $\mathcal{L}_n(\beta)$ and its impact to the rate of $\widehat{\beta}$. (**a**) Curved. (**b**) Flat

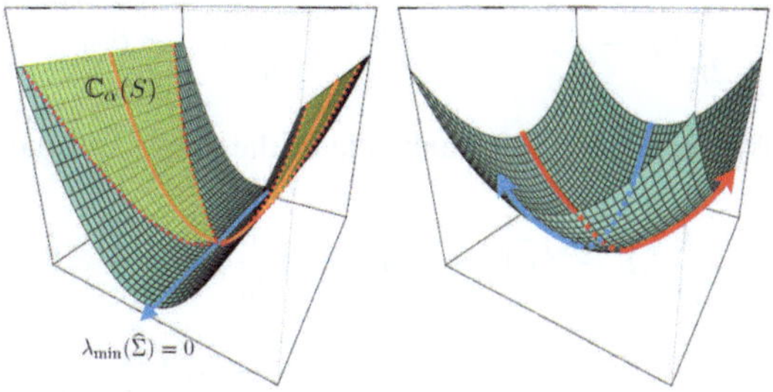

(a) (b)

Fig. 10.2 The surface of two types of Hessian matrices. (**a**) $\widehat{\Sigma} \succeq 0$. (**b**) $\widehat{\Sigma} \succ 0$

in the high dimensional setting, $\nabla^2 \mathcal{L}_n(\widehat{\beta}) = \mathbb{X}^\top \mathbb{X}/n = \widehat{\Sigma}$, whose minimum eigenvalue $\lambda_{\min}(\widehat{\Sigma}) = 0$. So the least squares loss is like Fig. 10.2a. In comparison, in the case when the Hessian is positive definite (see Fig. 10.2b), we can see that the curved space in Fig. 10.2a is flat along one direction and curved along another. But it is not the end of the world. Remember that β^* is sparse, so we do not need the loss to be curved along all directions, but only along some special directions related to the sparsity of β^*. Specifically, the RE condition assumes that the loss is curved along the cone $\mathbb{C}_\alpha(S)$.

10.2 Statistical Rate of Lasso

With the restricted eigenvalue condition, we can present the rate of Lasso estimator.

Theorem 10.1 (Rate of Lasso) *Let the cardinality of the support of β^* be $|S| = s$. Suppose $\mathbb{X}$ has* RE$(\kappa, 3)$. *If we choose the tuning parameter $\lambda \geq \frac{2}{n}\|\mathbb{X}^\top \varepsilon\|_\infty$, then*

$$\|\widehat{\beta} - \beta^*\|_2 \leq \frac{3}{\kappa}\sqrt{s}\lambda. \tag{10.3}$$

This theorem is totally deterministic: we do not impose any random assumption on the noise ε. On the right-hand side of (10.3), we can see that the rate of $\widehat{\beta}$ depends on κ, s and λ. The larger κ is, i.e., the more curved the loss is, the better is the rate. The rate also depends on the choice of the tuning parameter. This implies that as lambda increases, the rate will suffer as the estimator is more biased. The theorem suggests us to choose $\lambda = \frac{2}{n}\|\mathbb{X}^\top \varepsilon\|_\infty$; however, ε is unknown. We will handle this issue in the following corollary. Before that, we first present the proof of the theorem.

Proof In the first part of the proof, we aim to show that if $\lambda \geq \frac{2}{n}\|\mathbb{X}^\top \varepsilon\|_\infty$, then the error vector $\Delta = \widehat{\beta} - \beta^*$ has $\|\Delta_{S^c}\|_1 \leq 3\|\Delta_S\|_1$.

Our proof strategy is to begin with the zero-order optimization condition. Knowing the loss function is minimized for $\widehat{\beta}$, we have

$$\frac{1}{2n}\|Y-\mathbb{X}\widehat{\beta}\|_2^2+\lambda\|\widehat{\beta}\|_1 \le \frac{1}{2n}\|Y-\mathbb{X}\beta^*\|_2^2+\lambda\|\beta^*\|_1 = \frac{1}{2n}\|\varepsilon\|_2^2+\lambda\|\beta^*\|_1.$$

We expand the left-hand side of the above inequality as

$$\begin{aligned}\frac{1}{2n}\|Y-\mathbb{X}\widehat{\beta}\|_2^2+\lambda\|\widehat{\beta}\|_1 &= \frac{1}{2n}\|\mathbb{X}\beta^*-\mathbb{X}\widehat{\beta}+\|\varepsilon\|_2^2+\lambda\|\widehat{\beta}\|_1\\ &= \frac{1}{2n}\|\mathbb{X}\Delta\|_2^2-\left\langle\varepsilon,\frac{1}{n}\mathbb{X}^\top\Delta\right\rangle+\frac{1}{2n}\|\varepsilon\|_2^2+\lambda\|\widehat{\beta}\|_1.\end{aligned}$$

Rearranging the above two inequalities, we have

$$\begin{aligned}0\le\frac{1}{2n}\|\mathbb{X}\Delta\|_2^2 &\le \frac{\varepsilon^\top\mathbb{X}\Delta}{n}+\lambda\|\beta^*\|_1-\lambda\|\widehat{\beta}\|_1\\ &\le \frac{1}{n}\|\mathbb{X}^\top\varepsilon\|_\infty\|\Delta\|_1+\lambda\|\beta_S^*\|_1-\lambda\|\widehat{\beta}_S\|_1-\lambda\|\widehat{\beta}_{S^c}\|_1\\ &\le \frac{\lambda}{2}\|\Delta\|_1+\lambda\|\Delta_S\|_1-\lambda\|\Delta_{S^c}\|_1\\ &= \frac{\lambda}{2}(3\|\Delta_S\|_1-\|\Delta_{S^c}\|_1),\end{aligned}$$

where in the third inequality, we apply the triangle inequality $\|\beta_S^*\|_1-\|\widehat{\beta}_S\|_1\le\|\widehat{\beta}_S-\beta_S^*\|_1$ and $\Delta_{S^c}=\widehat{\beta}_{S^c}$ as $\beta_{S^c}^*=0$. Therefore, we have $\|\Delta_{S^c}\|_1\le3\|\Delta_S\|_1$ and thus $\Delta\in\mathbb{C}_3(S)$.

In the second part of the proof, we show the rate of Δ. As $\Delta\in\mathbb{C}_3(S)$, by $\mathrm{RE}(\kappa,3)$, $\frac{1}{n}\|\mathbb{X}\Delta\|_2^2\ge\kappa\|\Delta\|_2^2$. Going back to the inequality above, we have

$$\begin{aligned}\kappa\|\Delta\|_2^2 &\le \frac{1}{n}\|\mathbb{X}\Delta\|_2^2\le\lambda(3\|\Delta_S\|_1-\|\Delta_{S^c}\|_1)\le3\lambda\|\Delta_S\|_1\le3\lambda\sqrt{s}\|\Delta_S\|_2\\ &\le 3\lambda\sqrt{s}\|\Delta\|_2,\end{aligned}$$

where in the fourth inequality, we use $\|x\|_1\le\sqrt{d}\|x\|_2$ for $x\in\mathbb{R}^d$. Therefore, we have $\|\Delta\|_2\le\frac{3}{\kappa}\sqrt{s}\lambda$. □

Now, we discuss how to choose λ when the noise ε is random.

Corollary 10.1 *If the noises $\varepsilon_1,\ldots,\varepsilon_n$ are independent, and ε_i is sub-Gaussian with variance proxy σ^2 for all $i=1,\ldots,n$. The design matrix $\mathbb{X}$ is normalized such that the variance of the jth column of the design matrix $\mathbb{X}_j$ has $\frac{1}{n}\|\mathbb{X}_j\|_2^2\le1$ for all $1\le j\le d$. If $\mathbb{X}$ satisfies* $\mathrm{RE}(\kappa,3)$, *and we choose $\lambda=\sigma\sqrt{\log(2d/\delta)/(2n)}$, then with probability at least $1-\delta$, we have*

$$\|\widehat{\beta}-\beta^*\|_2\le\frac{3\sigma}{2\kappa}\sqrt{\frac{2s\log(2d/\delta)}{n}}.$$

This corollary implies that if we choose the tuning parameter $\lambda = C\sqrt{\log d/n}$ for some sufficiently large constant C, the Lasso estimator has the ℓ_2-norm rate

$$\|\widehat{\beta} - \beta^*\|_2 = O_P\left(\sqrt{\frac{s\log d}{n}}\right).$$

The Lasso estimator is consistent as long as $s\log d/n = o(1)$: if s is fixed, the dimension d can increase at the rate of the exponential of sample size. If we know the true support S and run the least squares $\widehat{\beta}^{\mathrm{LS}} = \arg\min_{\beta_S} \|Y - \mathbb{X}_S\beta_S\|_2^2$, Theorem 7.1 shows the rate of OLS is

$$\|\widehat{\beta}^{\mathrm{LS}} - \beta^*\|_2 = O_P(\sqrt{r/n}).$$

In comparison, the Lasso estimator has an additional $\sqrt{\log d}$ term in the rate, which is the price for variable selection since Lasso does not know the true support.

Proof By Theorem 10.1, we need to control $\|\mathbb{X}^\top\varepsilon\|_\infty/n$. We have

$$\mathbb{P}\left(\frac{1}{n}\|\mathbb{X}^\top\varepsilon\|_\infty > t\right) = \mathbb{P}\left(\max_{1\le j\le d}\left|\frac{1}{n}\mathbb{X}_j^\top\varepsilon\right| > t\right) \le d\mathbb{P}(|X_j^\top\varepsilon| > nt).$$

By Hoeffding inequality (Theorem 4.7), for any $j = 1, \ldots, d$, $\mathbb{X}_j^\top\varepsilon$ is sub-Gaussian with the variance proxy $\sigma^2\|\mathbb{X}_j\|_2^2 \le \sigma^2 n$. The sub-Gaussian tail gives us

$$\mathbb{P}\left(\frac{1}{n}\|\mathbb{X}^\top\varepsilon\|_\infty > t\right) \le d\mathbb{P}(|\mathbb{X}_j^\top\varepsilon| > nt) \le 2de^{-nt^2/(2\sigma^2)}.$$

Therefore, if we choose $\lambda = \sqrt{\frac{\sigma^2\log(2d/\delta)}{2n}}$, with probability at least $1-\delta$, we have $\lambda \ge \frac{2}{n}\|\mathbb{X}^\top\varepsilon\|_\infty$, and by Theorem 10.1,

$$\|\widehat{\beta} - \beta^*\|_2 \le \frac{3}{\kappa}\sqrt{s}\lambda = \frac{3}{\kappa}\sqrt{\frac{\sigma^2 s\log(2d/\delta)}{2n}}.$$

□

In the last part of the chapter, we give a concrete example of the design matrix $\mathbb{X}$ satisfying the restricted eigenvalue condition. Similar to how we construct $\mathbb{X}$ satisfying RIP in Theorem 9.2, we also generate $\mathbb{X}$ randomly.

Proposition 10.1 *Let the rows of $\mathbb{X}$: $X_1, \ldots, X_n$ be i.i.d. $N(0,\Sigma)$. We define the matrix maximum norm $\|A\|_{\max} = \max_{i,j}|A_{ij}|$. For any $\delta \in (0,1)$, if $64s\sqrt{\frac{\log(d/\delta)}{n}} \le \frac{\lambda_{\min}(\Sigma)}{\|\Sigma\|_{\max}}$, then*

$$\mathbb{P}\big(\mathbb{X} \textit{ satisfies } \mathrm{RE}(\lambda_{\min}(\Sigma)/2, 3)\big) \ge 1-\delta.$$

Therefore, if $s\sqrt{\log d/n} = o(1)$ and $X_i \overset{\text{iid}}{\sim} N(0, \Sigma)$, then $\mathbb{X}$ satisfies the RE condition with high probability. In fact, the scaling condition here is suboptimal. It is proved in [1] that the normal design is RE with high probability if $\sqrt{s \log d/n} = o(1)$.

Proof We first control the rate of the sample covariance matrix in $\|\cdot\|_{\max}$-norm. Since $X_i \sim N(0, \Sigma)$, for any $1 \le i, j \le d$, $X_{ij}X_{ik}$ is sub-exponential with the parameter $\Sigma_{jj}\Sigma_{kk} \le \|\Sigma\|_{\max}^2$. Therefore, we have

$$
\begin{aligned}
&\mathbb{P}\Big(\|\widehat{\Sigma} - \Sigma\|_{\max} > t\Big) \\
&\quad = \mathbb{P}\Big(\max_{1\le j\le k\le d} |\widehat{\Sigma}_{jk} - \Sigma_{jk}| > t\Big) \\
&\quad \le d^2\mathbb{P}\bigg(\Big|\frac{1}{n}\sum_{i=1}^{n}\Big(X_{ij}X_{ik} - \mathbb{E}[X_{ij}X_{ik}]\Big)\Big| > t\bigg) \qquad \text{(By union bound)} \\
&\quad \le d^2 e^{-\frac{nt^2}{2\|\Sigma\|_{\max}^2}}. \qquad \text{(By the concentration of sub-exponential)}
\end{aligned}
$$

Therefore, with probability at least $1 - \delta$,

$$
\|\widehat{\Sigma} - \Sigma\|_{\max} \le \|\Sigma\|_{\max}\sqrt{\frac{4\log(d/\delta)}{n}}.
$$

Second, we will show $\mathbb{X}$ satisfies RE. We have

$$
\begin{aligned}
\frac{1}{n}\|\mathbb{X}\Delta\|_2^2 &= \Delta^\top\widehat{\Sigma}\Delta = \Delta^\top(\widehat{\Sigma} - \Sigma)\Delta + \Delta^\top\Sigma\Delta \\
&\ge \lambda_{\min}(\Sigma)\|\Delta\|_2^2 - \|\widehat{\Sigma} - \Sigma\|_{\max}\|\Delta\|_1^2,
\end{aligned}
\tag{10.4}
$$

where, in the last inequality, we use (10.2) and the matrix Hölder inequality

$$
|x^\top Ax| \le \|x\|_1\|Ax\|_\infty \le \|A\|_{\max}\|x\|_1^2.
$$

When $\Delta \in \mathbb{C}_3(S)$, we have $\|\Delta\|_1 = \|\Delta_S\|_1 + \|\Delta_{S^c}\|_1 \le 4\|\Delta_S\|_1 \le 4\sqrt{s}\|\Delta\|_2$. Therefore, with probability at least $1 - \delta$,

$$
\|\widehat{\Sigma} - \Sigma\|_{\max}\|\Delta\|_1^2 \le 32\|\Sigma\|_{\max}\sqrt{\frac{s^2\log(d/\delta)}{n}}\|\Delta\|_2^2 \le \frac{\lambda_{\min}(\Sigma)}{2}\|\Delta\|_2^2,
$$

as we assume $64s\sqrt{\frac{\log(d/\delta)}{n}} \le \frac{\lambda_{\min}(\Sigma)}{\|\Sigma\|_{\max}}$. Combining with (10.4), we have

$$
\frac{1}{n}\|\mathbb{X}\Delta\|_2^2 \ge \frac{1}{2}\lambda_{\min}(\Sigma)\|\Delta\|_2^2.
$$

□

Chapter 11
Variations of Lasso

11.1 Limitations of Lasso

In the previous chapter, we study the high-dimensional linear model $Y = \mathbb{X}\beta^* + \epsilon$, with $\mathbb{X} \in \mathbb{R}^{n\times d}$ and $\|\beta^*\|_0 \le s$. We propose to estimate β^* via Lasso estimator

$$\widehat{\beta}^{\text{Lasso}} = \arg\min_{\beta} \frac{1}{2n}\|Y - \mathbb{X}\beta\|_2^2 + \lambda\|\beta\|_1.$$

We consider two assumptions: (1) the design matrix satisfies the restricted eigenvalue condition and (2) the noises ε are independent sub-Gaussians with variance proxy σ^2. If we choose $\lambda = C\sigma\sqrt{\log d/n}$ for some sufficiently large constant C, we show that the Lasso estimator has the statistical rate $\|\widehat{\beta}^{\text{Lasso}} - \beta^*\|_2 = O_P(\sqrt{s\log d/n})$.

However, the Lasso estimator is not perfect. In this chapter, we will discuss the extensions of Lasso. Due to the scope of our book, we could not cover all details of these extensions. In order to extend Lasso, we first list the limitations of Lasso below. We will elaborate these limitations in the remaining of this chapter.

Limitations of Lasso

1. **Model**: Lasso is restricted to the linear model.
2. **Penalty**: The ℓ_1-penalty of Lasso can only regularize the sparsity.
3. **Bias**: Lasso estimator is biased.
4. **Tuning**: The choice of λ depends on the unknown σ.
5. **Robustness**: Lasso is consistent only when ε has light tail.

J. Lu, *Big Data Analysis*, https://doi.org/10.1007/978-3-032-03161-7_11

11.2 Beyond Linear Model

We first generalize the Lasso to general high-dimensional models. In general, we observe the samples $(Y_1, X_1), \ldots, (Y_n, X_n)$ i.i.d. generated from the density $p_{\beta^*}(y, x)$. If the dimension of the parameter $\beta^* \in \mathbb{R}^d$ is sparse, we can estimate β^* via adding an ℓ_1-penalty to the log-likelihood. Therefore, we have the following general Lasso estimation:

$$\min_{\beta} \sum_{i=1}^{n} - \log p_{\beta}(Y_i, X_i) + \lambda \|\beta\|_1.$$

In fixed dimensional statistics, besides the maximal likelihood estimator, we may also consider some other loss between Y_i and X_i's, denoted as $\mathcal{L}_n(\beta)$. In principle, we can transform them to estimate sparse parameters by considering the problem $\min_\beta \mathcal{L}_n(\beta) + \lambda\|\beta\|_1$. We will give a few concrete examples below.

11.2.1 High-Dimensional Classification

Let us start with the classification problem. Comparing to the regression problem, the classification has the binary response $Y \in \{-1, +1\}$. The arguably most popular model in classification is the logistic model. The **linear logistic model** assumes

$$\mathbb{P}(Y = +1|X = x) = \frac{1}{1 + e^{-x^\top \beta^*}} \text{ and } \mathbb{P}(Y = -1|X = x) = \frac{1}{1 + e^{x^\top \beta^*}}.$$

If we observe i.i.d. samples $(Y_1, X_1), \ldots, (Y_n, X_n)$ from the linear logistic model and β^* is sparse, we can estimate β^* via adding the ℓ_1-penalty to the logistic loss as follows.

The ℓ_1**-penalized logistic regression**:

$$\min_{\beta} \sum_{i=1}^{n} \log\left(1 + \exp(-Y_i X_i^\top \beta)\right) + \lambda\|\beta\|_1.$$

We now switch to another widely used classification model. In logistic model, we impose the model on $\mathbb{P}(Y|X)$. We can also turn to model $\mathbb{P}(X|Y)$. Specifically, the **discriminative classification model** assumes the data (Y, X) generated from the above model follows the mechanism:

1. Generate a Bernoulli random variable $Y \sim \text{Bernoulli}(\eta)$.
2. If $Y = +1$, generate $X \sim p_+(x)$; otherwise, generate $X \sim p_-(x)$.

This model is also called the **mixture model** as it is a mixture of two distributions.

The **Gaussian discriminant model** further assumes that p_+ and p_- are Gaussian distributions. Specifically, we denote

$$p_+(x) = \frac{1}{(2\pi)^{\frac{d}{2}}|\Sigma|^{\frac{1}{2}}} \exp\left(-\frac{1}{2}(x-\mu_+)^\top \Sigma^{-1}(x-\mu_+)\right);$$

$$p_-(x) = \frac{1}{(2\pi)^{\frac{d}{2}}|\Sigma|^{\frac{1}{2}}} \exp\left(-\frac{1}{2}(x-\mu_-)^\top \Sigma^{-1}(x-\mu_-)\right).$$

Given an input $X = x$, we can use the following Bayes rule to decide whether the corresponding Y is $+1$ or -1:

$$\text{Bayes rule: } h(x) = \begin{cases} +1 \text{ if } \mathbb{P}(Y=+1|X=x) \geq \mathbb{P}(Y=-1|X=x); \\ -1 \text{ otherwise.} \end{cases}$$

We can see that the Bayes rule classifies Y as $+1$ as long as the condition probability of $Y = +1$ with the prior information of $X = x$ is larger than the probability that $Y = -1$. For the discriminative classification model, we can get the closed forms of these two probabilities in the Bayes rule:

$$\mathbb{P}(Y=+1|X=x) = \frac{p_+(x)\eta}{p_+(x)\eta + p_-(x)(1-\eta)},$$

$$\mathbb{P}(Y=-1|X=x) = \frac{p_-(x)(1-\eta)}{p_+(x)\eta + p_-(x)(1-\eta)}.$$

Therefore, we can reformulate the Bayes rule as

$$\log\frac{\mathbb{P}(Y=+1|X=x)}{\mathbb{P}(Y=-1|X=x)} = \log\frac{p_+(x)}{p_-(x)} + \log\frac{\eta}{1-\eta} \geq 0.$$

Plugging in the Gaussian densities, under the balanced setting, i.e., $\eta = 1/2$, the Bayes rule becomes

$$f(x) = \underbrace{(\mu_+ - \mu_-)^\top \Sigma^{-1}}_{\beta^{*\top}} \Big(x - \underbrace{\frac{\mu_+ + \mu_-}{2}}_{\bar{\mu}}\Big) \geq 0,$$

where we can see that the decision rule is linear. Therefore, we call this method as **linear discriminant analysis** (LDA). When $\eta \neq 1/2$, it is easy to find that the decision boundary is also linear.

Suppose $(Y_1, X_1), \ldots, (Y_n, X_n)$ are i.i.d. from the Gaussian discriminant model. In order to get the decision boundary $f(x) = \beta^{*\top}(x - \bar{\mu})$, we need to estimate

$$\beta^* = \Sigma^{-1}(\mu_+ - \mu_-) \text{ and } \bar{\mu} = \frac{\mu_+ + \mu_-}{2}.$$

In high-dimensional LDA, we assume that β^* is sparse as the parsimonious principle implies that only a few variables among d covariates will be involved in the decision rule. We start with estimating the mean and covariance matrices as

$$\begin{aligned}
n_+ &= \textstyle\sum_{i=1}^n \mathbb{I}(Y_i = +1), & n_- &= \textstyle\sum_{i=1}^n \mathbb{I}(Y_i = -1);\\
\widehat{\mu}_+ &= \tfrac{1}{n_+}\textstyle\sum_{i:Y_i=+1} X_i, & \widehat{\mu}_- &= \tfrac{1}{n_-}\textstyle\sum_{i:Y_i=-1} X_i;\\
\widehat{\Sigma}_+ &= \tfrac{1}{n_+}\textstyle\sum_{i:Y_i=+1}(X_i - \widehat{\mu}_+)(X_i - \widehat{\mu}_+)^\top, & \widehat{\Sigma}_- &= \tfrac{1}{n_-}\textstyle\sum_{i:Y_i=-1}(X_i - \widehat{\mu}_-)\\
& & &\quad (X_i - \widehat{\mu}_-)^\top.
\end{aligned}$$

We can then estimate $\bar{\mu}$ and the covariance matrix estimator as

$$\widehat{\bar{\mu}} = \frac{\widehat{\mu}_+ + \widehat{\mu}_-}{2} \text{ and } \widehat{\Sigma} = \frac{n_+\widehat{\Sigma}_+ + n_-\widehat{\Sigma}_-}{n_+ + n_-}.$$

Notice that Σ^{-1} is involved in the definition of β^*. We know that in the high-dimensional setting, $\widehat{\Sigma}$ is not invertible. However, we observe that $\Sigma\beta^* = \mu_+ - \mu_-$. Therefore, we would expect $\widehat{\Sigma}\beta^*$ is close to $\widehat{\mu}_+ - \widehat{\mu}_-$. This motivates us to consider the following high-dimensional LDA.

The **high-dimensional linear discriminant analysis**:

$$\widehat{\beta}^{\text{LDA}} = \arg\min_{\beta} \|\beta\|_1, \text{ s.t. } \|\widehat{\Sigma}\beta - (\widehat{\mu}_+ - \widehat{\mu}_-)\|_\infty \le \lambda,$$

and the decision rule is $f(x) = (x - \widehat{\bar{\mu}})^\top \widehat{\beta}^{\text{LDA}} \ge 0$. The problem above can be solved via linear programming.

11.2.2 High-Dimensional Graphical Model

Networks are widely used as an interpretable way to visualize the data. An undirected graph $G = (V, E)$ has the vertex set V and edge set E. We denote the number of nodes $d = |V|$. We propose the following graphical model connecting the graph structure with the conditional dependence.

Definition 11.1 (Graphical Model) A random vector $X \in \mathbb{R}^d$ is Markov with respect to a graph $G = (V, E)$ if

$$(j, k) \notin E \Leftrightarrow X_j \perp\!\!\!\perp X_k | \text{ other } X_i\text{'s} .$$

By the definition, the node j is connected to node k if X_j are dependent to X_k conditioning the rest $\{X_\ell | \ell \neq j, k\}$. The graphical model is widely used in real applications. In practice, X could be gene expressions, brain imaging, or social media. The graph G corresponding to X then becomes the genomic network, brain network, or social network. Comparing to the unconditional dependency like correlation, the conditional dependency can characterize direct connections between variables.

For the normal distributions, the following proposition implies that we can characterize the graphical model directly with the inverse covariance matrix.

Proposition 11.1 (Gaussian Graphical Model) *If $X \sim N(0, \Sigma)$, then X is Markov to $G = (V, E)$ if*

$$(j, k) \in E \Leftrightarrow (\Sigma^{-1})_{jk} \neq 0.$$

This proposition says the edges of Gaussian graphical model are determined by the inverse covariance matrix $\Theta = \Sigma^{-1}$, which is also called the precision matrix.

A sparse graphical model assumes that the graph G is sparse, i.e., the number of edges is much smaller than the number of nodes d. For the Gaussian graphical model, this is to assume that the precision matrix is sparse. Let $X_1, \ldots, X_n$ be i.i.d. samples from $N(0, \Sigma)$. Under the high-dimensional setting, we cannot estimate the precision matrix via directly inverting the sample covariance matrix $\widehat{\Sigma}$ as it is not invertible. However, we can combine the negative log-likelihood of $N(0, \Sigma)$ with the ℓ_1-penalty.

The **graphical Lasso** (GLasso) estimator is

$$\min_{\Theta \succeq 0} \mathrm{Tr}(\widehat{\Sigma}\Theta) - \log\det(\Theta) + \lambda||\Theta||_{1,1},$$

where the matrix ℓ_1-norm is $||\Theta||_{1,1} = \sum_{j \neq k} |\Theta_{jk}|$.

11.3 Beyond the ℓ_1-Penalty

Sparsity plays a vital role in the high-dimensional statistics. We know that the ℓ_1-norm $||\beta||_1$ encourages sparsity of β. However, we may need to impose other structural assumptions on the parameter β. Then we need to consider other penalty terms on β.

11.3.1 Group Lasso

The Lasso estimator can be used to select variables from the high-dimensional features. Sometimes the features are in groups, and we want to select variables in groups.

Suppose $\beta \in \mathbb{R}^d$ has J groups. We denote each group as $S_j \subset \{1, \ldots, d\}$ for $j = 1, \ldots, J$. Therefore, we want to select sub-vectors $\beta_{S_1}, \ldots, \beta_{S_J}$.

For example, we want to predict the number of COVID cases tomorrow. The covariates to predict Y are groups: (1) the group of features related to the number of cases in the past (the number of cases today, the number of cases yesterday, the number of cases in the past month, etc.), (2) the group of features related to the weather (temperature, precipitation, and so on), (3) the group of features related to the social distancing (the number of people working from home, the number of restaurants opened, and so on), and (4) the group of features related to Trump's activities (the number of Trump's tweets, the number of days Trump recovered from COVID, and so on). We may expect that the number of COVID cases is related to some group of these variables.

If β is sparse in groups, we have the vector $(||\beta_{S_1}||_2, ||\beta_{S_2}||_2, \ldots, ||\beta_{S_J}||_2)^\top \in \mathbb{R}^J$ is sparse. Therefore, we consider the group Lasso penalty as

$$\left\|(||\beta_{S_1}||_2, ||\beta_{S_2}||_2, \ldots, ||\beta_{S_J}||_2)\right\|_1 = \sum_{j=1}^{J} ||\beta_{S_j}||_2.$$

Therefore, the group Lasso estimator is formulated as follows.

The **group Lasso** estimator:

$$\min_{\beta} \left\| Y - \sum_{j=1}^{J} \mathbb{X}_{S_j}\beta_{S_j} \right\|_2^2 + \lambda \sum_{j=1}^{J} ||\beta_{S_j}||_2.$$

An important application of the group Lasso is to the **sparse additive model** (SPAM):

$$Y_i = \sum_{j=1}^{d} f_j(X_{ij}) + \varepsilon_i, \quad \text{for } i = 1, \ldots, n,$$

where only s of these univariate functions f_j's is nonzero. To estimate f_j's, we expand the function in basis:

$$f_j(x_j) = \sum_{k=1}^{\infty} \beta^*_{jk}\phi_k(x_j), \text{ for } j = 1, \ldots, d,$$

where $\{\phi_k\}_{k=1}^{\infty}$ are basis functions, e.g., the polynomial bases $\{x^k\}_{k=1}^{\infty}$, the trigonometric bases $\{\sin(kx), \cos(kx)\}_{k=1}^{\infty}$, the B-splines, etc.

Therefore, if we want to select the nonparametric functions f_j's, it is equivalent to select the basis coefficients $\{\beta^*_{jk}\}_{k=1}^{\infty}$ for $j = 1, \ldots, d$ in group. This motivates us to consider the sparse additive model estimator with group Lasso penalty as follows.

The **spare additive model estimator**:

$$\min_{\beta_{jk}} \frac{1}{n}\sum_{i=1}^{n}\Big(Y_i - \sum_{j=1}^{d}\sum_{k=1}^{m}\beta_{jk}\phi_k(X_{ij})\Big)^2 + \lambda\sum_{j=1}^{d}\Big(\sum_{k=1}^{m}\beta_{jk}^2\Big)^{\frac{1}{2}},$$

where m is the number of basis functions we choose to approximate the true function.

11.3.2 Fused Lasso

In many applications, the parameters β are not sparse but piecewise constant. For example, for the genomic data in Fig. 11.1, the expression levels of neighbor genes are the same.

Although β is not sparse, but if β is piecewise constant, then the differences $(\beta_1 - \beta_2, \beta_2 - \beta_3, \ldots, \beta_{d-1} - \beta_d)^\top = D\beta$, where D is the differential map

$$D = \begin{pmatrix} 1 & -1 & 0 & \cdots & 0 & 0 \\ 0 & 1 & -1 & \cdots & 0 & 0 \\ \vdots & \vdots & \vdots & \ddots & \vdots & \vdots \\ 0 & 0 & 0 & \cdots & 1 & -1 \end{pmatrix} \in \mathbb{R}^{(d-1)\times d},$$

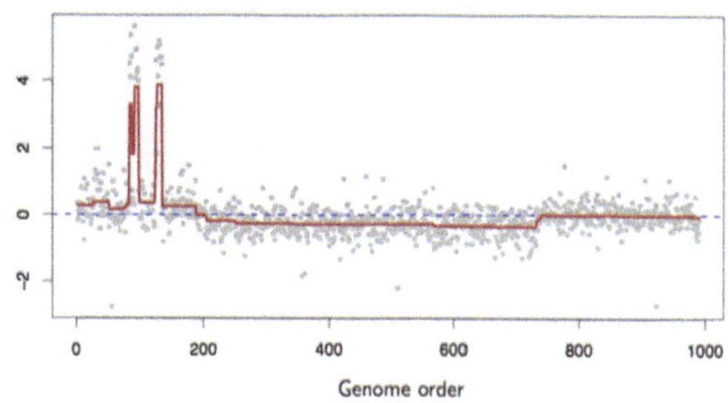

Fig. 11.1 The genomic sequencing data (gray dots) and the truth (red lines). We can see that the signal is a mixture of spiked and piecewise constant signals

are sparse. Therefore, instead of penalizing $\|\beta\|_1$, we should penalize the ℓ_1-norm of $D\beta$. This gives us the fused Lasso estimator.

The **fused Lasso** estimator:

$$\min_{\beta} ||Y - \mathbb{X}\beta||_2^2 + \lambda||D\beta||_1.$$

We can also apply the fused Lasso to 2D image data. For example, in Fig. 11.2, β^* is the brain imaging data and Y is a noisy observation. We can see that the β is a "patched" image: it is a two-dimensional piecewise constant. Therefore, we can apply the differential map D to the rows and columns of β and expect that both of them are sparse. Then we can recover the truth from the noisy observation via

$$\min_{\beta} \sum_{jk} |Y_{jk} - \beta_{jk}|^2 + \lambda||D\beta||_1 + \lambda||D\beta^\top||_1.$$

11.4 Beyond the Biasedness

Now we go back to the linear model $Y = \mathbb{X}\beta^* + \varepsilon$ with the sparsity assumption. Even if β is sparse, the ℓ_1-norm may not be perfect. Ideally, we want to regularize the ℓ_0-norm of β, but we know that the computation of ℓ_0-norm is challenging. That is why we introduce the convex relaxation of the ℓ_0-norm, which is ℓ_1-norm. Despite its advantage in computation, the larger β_j is, the more the ℓ_1-norm penalizes it, while the ℓ_0-norm does not. Therefore, Lasso is biased for large β_j's.

A straightforward approach to fix this problem is through refitting. Let $\widehat{S}$ be the support of $\widehat{\beta}^{\text{Lasso}}$, and we can refit the linear model on $\widehat{S}$ via the least squares. Denote the refitted least squares estimator as

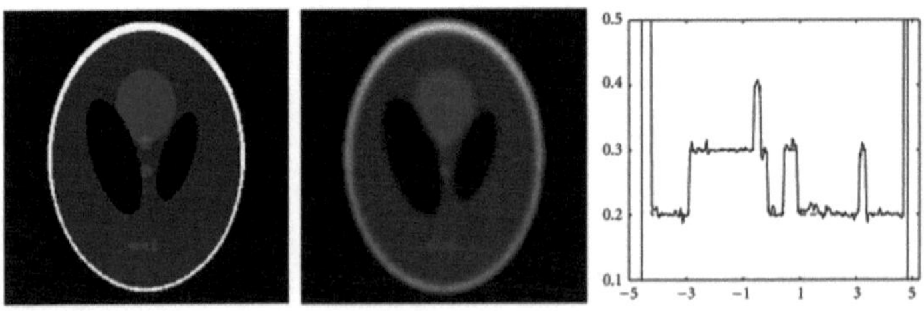

(a) Truth (b) Noisy (c) Sliced Signal

Fig. 11.2 Shepp-Logan phantom image. Panel (**a**) is the original signal, panel (**b**) is the noisy image, and panel (**c**) is one row of the image

$$\widehat{\beta}^{\mathrm{LS}}_{\widehat{S}} = (\mathbb{X}_{\widehat{S}}^{\top}\mathbb{X}_{\widehat{S}})^{\dagger}\mathbb{X}_{\widehat{S}}^{\top}Y \text{ and } \widehat{\beta} = (\widehat{\beta}_{\widehat{S}}^{\top}, \widehat{\beta}_{\widehat{S}^c}^{\top})^{\top}, \text{ where } \widehat{\beta}_{\widehat{S}} = \widehat{\beta}^{\mathrm{LS}}_{\widehat{S}}, \widehat{\beta}_{\widehat{S}^c} = 0.$$

Since the least squares is unbiased, so the refitted estimator should remove the bias of Lasso estimator.

Besides the refitted estimator, the adaptive Lasso fixes the bias. The idea of the adaptive Lasso is that if β_j^* is large, we should penalize less on β_j. Let $\widehat{\beta}^{\mathrm{ini}}$ be some initial estimator of β^*, which could be Lasso estimator or others.

The **adaptive Lasso** estimator is

$$\min_{\beta} \|Y - \mathbb{X}\beta\|_2^2 + \lambda \sum_{j=1}^{d} \frac{|\beta_j|}{|\widehat{\beta}_j^{\mathrm{ini}}|^{\gamma}}, \quad \text{where } \gamma > 0.$$

We can see that the adaptive Lasso has a weighted ℓ_1-norm as the penalty. The larger $|\widehat{\beta}_j^{\mathrm{ini}}|$ is, the weight on $|\beta_j|$ is smaller, and thus the bias is smaller.

11.4.1 Nonconvex Penalties

The previous two approaches fixing the bias of Lasso depend on some initial estimator. In fact, we can formulate a one-step estimator. We can see that the ℓ_1-norm is convex but biased while ℓ_0 is discontinuous but unbiased. As a trade-off, we can consider a penalty which is between ℓ_1 and ℓ_0. The SCAD (Smoothly Clipped Absolute Deviation) penalty is this kind of penalty.

The SCAD penalty is

$$p_{\lambda}^{\mathrm{SCAD}}(t) = \int_0^t \Big(\mathbb{I}\{t \le \lambda\} + \frac{a\lambda - x}{(a-1)\lambda}\mathbb{I}\{t > \lambda\}\Big)dx,$$

and the SCAD estimator is

$$\widehat{\beta}^{\mathrm{SCAD}} = \arg\min_{\beta} \|Y - \mathbb{X}\beta\|_2^2 + 2\sum_{j=1}^{d} p_{\lambda}^{\mathrm{SCAD}}(\beta_j).$$

When $\mathbb{X}$ is identity, the penalized least squares $\min_{\beta} \|Y - \beta\|_2^2 + p_{\lambda}(\beta)$ has closed form solution for the penalty p_{λ} to be ℓ_0, ℓ_1, or SCAD.

We can check that SCAD penalty is nonconvex. However, there are efficient algorithms to solve SCAD estimator. Denote two index sets:

- Strong signals: $S_1 = \{j \in S \| |\beta_j^*| \geq C\sqrt{\log d/n}\}$ for some sufficiently large constant C.
- Weak signals: $S_2 = S \backslash S_1$.

It can be shown that, if we choose the tuning parameter properly, the SCAD estimator has the statistical rate

$$||\widehat{\beta}^{\text{SCAD}} - \beta^*||_2 = O_P\left(\sqrt{\frac{|S_1|}{n}} + \sqrt{\frac{|S_2|\log d}{n}}\right).$$

In comparison, the Lasso estimator has the rate (Corollary 10.1)

$$||\widehat{\beta}^{\text{Lasso}} - \beta^*||_2 = O_P\left(\sqrt{\frac{(|S_1| + |S_2|)\log d}{n}}\right).$$

As we discussed in the previous chapter, the $\sqrt{\log d}$ term in the rate can be treated as a price to select s variables from the d features. The statistical rate of SCAD estimator shows that SCAD does not need to pay the $\sqrt{\log d}$ price to detect strong signals, which illustrates how SCAD does not have bias for strong enough signals in theory.

11.5 Beyond Tuning Sensitive

Recall that in Corollary 10.1, if the noises ε are independent sub-Gaussians with variance proxy σ^2, we suggest choosing the tuning parameter of Lasso as $\lambda = C\sigma\sqrt{\log d/n}$ for some sufficiently large constant C. However, in practice, we do not know the variance proxy σ^2. Although we can always use cross-validation in practice, the tuning procedure will be sensitive. We will introduce a tuning-insensitive variation of Lasso.

11.5.1 *Square-Root Lasso*

Intuitively, the reason why Lasso is tuning sensitive of the parameter of ε is that the gradient of the least squares loss $\mathcal{L}_n(\beta) = \|Y - \mathbb{X}\beta\|_2^2$ at the truth is $\nabla\mathcal{L}_n(\beta^*) = \mathbb{X}^\top\varepsilon$, which is related to ε. This is why in Theorem 10.1, we suggest to choose $\lambda \geq \frac{2}{n}\|\mathbb{X}^\top\varepsilon\|_\infty$. If we can design some loss whose gradient is irrelevant to ε, then we can expect the estimator to be tuning insensitive. A candidate loss is the square-root loss $\mathcal{L}_n^{1/2}(\beta) = \|Y - \mathbb{X}\beta\|_2$ whose gradient at the truth is

$$\nabla \mathcal{L}_n^{1/2}(\beta^*) = \frac{\mathbb{X}^\top \varepsilon}{2\|Y - \mathbb{X}\beta^*\|_2} = \frac{\mathbb{X}^\top \varepsilon}{2\|\varepsilon\|_2}.$$

Although it is still related to ε, we can see that the gradient of the square-root loss is somehow normalized and thus insensitive to the magnitude of ε. This motivates us to consider the square-root Lasso.

The **square-root Lasso** estimator is

$$\widehat{\beta}^{\sqrt{\text{Lasso}}} = \arg\min_{\beta} \frac{1}{\sqrt{n}} ||Y - \mathbb{X}\beta||_2 + \lambda\|\beta\|_1.$$

Under mild regularity conditions on the noise ε, it can be shown that if we choose $\lambda = C\sqrt{\log d/n}$ for some universal constant C irrelevant to ε, the square-root Lasso has the rate $\|\widehat{\beta}^{\sqrt{\text{Lasso}}} - \beta^*||_2 = O_P(\sqrt{s \log d/n})$.

11.6 Beyond Sub-Gaussian

Recall that in Corollary 10.1, the Lasso has the rate $O_P(\sqrt{s \log d/n})$ if the noises ε are independent sub-Gaussians. We aim to generalize the estimator to heavy-tailed noises. If ε is heavy-tailed, its moment-generating function may be infinite, but the quantile of ϵ is always finite. This motivates us to consider utilizing the property of quantiles.

11.6.1 Quantile Regression

We start with formulating the quantile as an optimization problem.

Proposition 11.2 (Variational Form of Quantiles) *Let $\rho_\tau(y) = y(\tau - \mathbb{I}\{y < 0\})$. Then τ-quantile of a random variable $X = \arg\min_x \mathbb{E}[\rho_\tau(X - x)]$. Specifically, we have*

$$\textit{Median of } X = \arg\min_{x} \mathbb{E}|X - x|.$$

Proof We denote $x_\tau = \arg\min_x \mathbb{E}[\rho_\tau(X - x)]$. Therefore, we have

$$0 = \frac{d}{dx}\mathbb{E}[\rho_\tau(X - x)]\Big|_{x=x_\tau} = \tau - \mathbb{E}[\mathbb{I}\{X - x_\tau < 0\}].$$

Thus, $\mathbb{P}(X < x_\tau) = \tau$, i.e., x_τ is the τ-quantile of X. □

Therefore, to have a robust estimator for the linear model, we can replace the least squares loss in Lasso by the $\rho_\tau(y) = y(\tau - \mathbb{I}\{y < 0\})$.

The **high-dimensional quantile regression** estimator is

$$\min_\beta \sum_{i=1}^{n} \rho_\tau(Y_i - X_i^\top \beta) + \lambda||\beta||_1.$$

When $\tau = 0.5$, it reduces to the **least absolute deviation estimator** (LAD):

$$\min_\beta \|Y - \mathbb{X}\beta\|_1 + \lambda||\beta||_1.$$

Since $\rho_\tau(\cdot)$ is piecewise linear, so the quantile regression can be solved via linear programming.

Part IV
High-Dimensional Optimization

Chapter 12
Convexity and Subgradient

12.1 Convex Optimization

From the previous chapters, we can see that many estimators can be formulated as an optimization problem. One example is the Lasso estimator

$$\min_{\beta \in \mathbb{R}^d} \frac{1}{2n} \|Y - \mathbb{X}\beta\|_2^2 + \lambda \|\beta\|_1.$$

The Lasso estimator is an unconstrained problem as we find the minimizer over the entire $\mathbb{R}^d$. In comparison, the constrained Lasso

$$\min_{\beta} \|Y - \mathbb{X}\beta\|_2^2, \text{ s.t. } \|\beta\|_1 \leq t,$$

has the constraint $\|\beta\|_1 \leq t$. More examples can be found in the examples of Chap. 11. Therefore, the next part of our class will focus on how to solve these optimization problems and how well these algorithms perform.

Under the setting of high-dimensional statistics, both the sample size n and the feature dimension d are very large. We talk about how to handle this challenge in probability and statistics. The big data challenge is also crucial in optimization from two perspectives: storage and computation. Given a d-dimensional function $f : \mathbb{R}^d \to \mathbb{R}$, in optimization, we will usually use its gradient $\nabla f \in \mathbb{R}^d$ and Hessian matrix $\nabla^2 f \in \mathbb{R}^{d \times d}$. In high-dimensional optimization, we will try to only use the gradients and avoid using the Hessian matrix as it is much slower to conduct matrix operations than vector operations. Moreover, when d is ultra-large, we may not even afford to store the d by d matrix, not mentioning the matrix operations. For example, many genomic datasets have expression levels of millions of genes; then the size of Hessian matrix related to the question will be trillions.

The optimization algorithms only involving the gradients are called the **first-order method**. Therefore, in our class, we will focus on the first-order method.

J. Lu, *Big Data Analysis*, https://doi.org/10.1007/978-3-032-03161-7_12

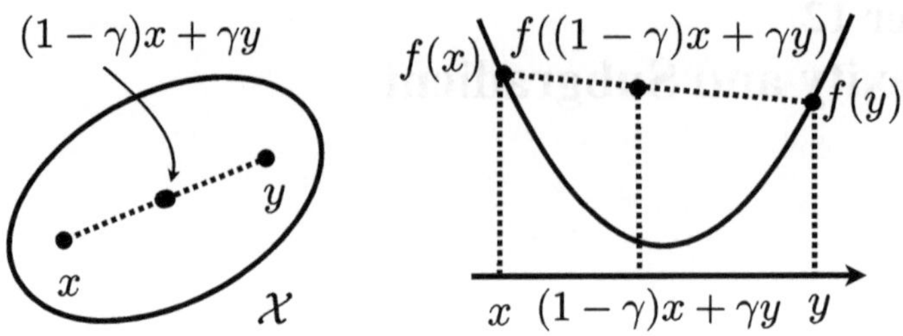

(a) (b)

Fig. 12.1 A set M is convex if it contains all its segments and a function f is convex if it always lies below its chords. (**a**) Convex set. (**b**) Convex function

From the unconstrained and constrained Lasso above, we found that both the least squares loss $\|Y - \mathbb{X}\beta\|_2^2$ and the ℓ_1-ball $\|\beta\|_1 \leq t$ are convex. Convexity is a very important concept in optimization. It covers a lot of useful examples and will induce many nice properties.

Let us start with formally defining a convex set and convex function.

Definition 12.1 (Convex Set and Function) A set M is convex if

$$(1-\gamma)x + \gamma y \in M, \text{ for any } x, y \in M, \gamma \in [0, 1].$$

A function $f : M \to \mathbb{R}$ is convex if

$$f((1-\gamma)x + \gamma y) \leq (1-\gamma)f(x) + \gamma f(y), \text{ for any } x, y \in M, \gamma \in [0, 1].$$

From Fig. 12.1, we can see that a set M is convex if it contains all its segments and a function f is convex if it always lies below its chords.

We next give a general form of optimization problems we are interested in our class.

Definition 12.2 (Convex Optimization) An optimization problem is called convex optimization if it can be formulated as

$$\min_x f(x)$$

$$\text{s.t. } x \in M,$$

where both f and M are convex.

We can see that both the unconstrained and constrained Lasso are convex optimizations. We can check that all the examples given by Chap. 11 (except SCAD) are convex optimizations as well.

12.2 Subgradient

As we like the first-order method in high-dimensional optimization, we need to study the properties of gradients. However, many convex functions are not smooth and do not have gradients. For example, the ℓ_1-norm $\|x\|_1$ is not differentiable at $x = 0$. Therefore, we will generalize the concept of gradient to non-smooth convex functions.

Definition 12.3 (Subgradient) Let $f : M \to \mathbb{R}$. We say $g \in \mathbb{R}^d$ is a subgradient of f at x if for any $y \in M$, we have

$$f(y) - f(x) \geq g^\top (y - x).$$

We denote the set of all subgradients of f at x as $\partial f(x)$.

Figure 12.2 illustrates the geometric interpretation of the subgradient. The hyperplane tangent to f at x is $p(y) = g^\top (y - x) + f(x)$. We can see in Fig. 12.2 that the inequality $f(y) - f(x) \geq g^\top (y - x)$ implies that f is above the hyperplane. So we also call $p(y)$ as a supporting hyperplane.

Notice that $\partial f(x)$ is a set. It can be shown (or seen from Fig. 12.2) that when a function is convex, the subgradient must exist. When f is differentiable and convex, by Definition 12.1, we have

$$\begin{aligned} f(y) &\geq \frac{f((1-\gamma)x + \gamma y) - (1-\gamma) f(x)}{\gamma} \\ &= f(x) + \frac{f(x + \gamma(y - x)) - f(x)}{\gamma} \xrightarrow[\gamma \to 0]{} f(x) + \nabla f(x)^\top (y - x). \end{aligned}$$

So $\nabla f(x) \in \partial f(x)$. However, the following example shows that the subgradient may be nonunique.

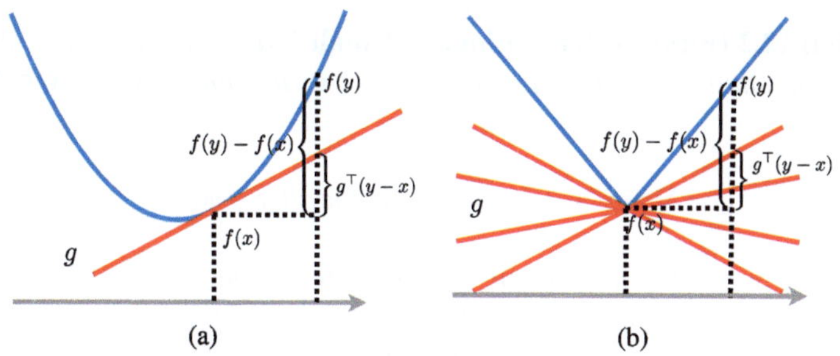

Fig. 12.2 The geometry of the subgradient. For smooth functions, the subgradient is unique but for non-smooth functions, the subgradient may be nonunique. (**a**) Smooth function. (**b**) Non-smooth function

Example 12.1 (Subgradient of ℓ_1-Norm) The most widely used non-smooth convex function in our class is the ℓ_1-norm. We first start with the univariate case. Let $f(x) = |x|$, and we know f is differentiable when $x \neq 0$. Therefore, $\partial|x| = \text{sign}(x)$ when $x \neq 0$. When $x = 0$, $|x|$ is non-smooth. From Fig. 12.2b, we can see that there are multiple lines which can "support" $|x|$. It is easy to check that for any $g \in [-1, 1]$, we have $|y| \geq gy$ for any y. Therefore, $\partial|x| = [-1, 1]$ when $x = 0$. We can generalize to the ℓ_1-norm $\|x\|_1$. Following the same analysis of the univariate case, if $g \in \partial\|x\|_1$, $g_j = \text{sign}(x_j)$ for $x_j \neq 0$ and $g_j \in [-1, 1]$ for $x_j = 0$.

We mentioned above that convexity will give us many nice properties. The following proposition is one of the reason why convex optimization is much easier than nonconvex optimization.

Proposition 12.1 (Local Minima Are Global Minima) *Let f be convex. If x^* is local minimum of f, then x^* is its global minimum. This happens if and only if $0 \in \partial f(x^*)$.*

Proof We say x^* is local minimum of f if, for any y, there exists a small enough γ such that

$$f(x^*) \leq f((1-\gamma)x^* + \gamma y) \leq (1-\gamma)f(x^*) + \gamma f(y),$$

where the last inequality is due to the convexity of f. Therefore, $f(x^*) \leq f(y)$ and x^* is global minimum.

If x^* is global minimum, $f(y) - f(x^*) \geq 0$ for any y. So $0 \in \partial f(x^*)$ by definition. It is straightforward to check the reverse direction. □

If x^* is global minimum of f if and only if $0 \in \partial f(x^*)$. We also call $0 \in \partial f(x^*)$ as the **first-order optimality condition**. Compared to the **zero-order optimality condition** $f(x^*) \leq f(y)$ for any y, the first-order optimality condition is called first as it involves the derivatives. Notice that $0 \in \partial f(x^*)$ is the first-order optimality condition only for the unconstrained optimization. The following proposition gives us the optimality condition for unconstrained problem.

Proposition 12.2 (First-Order Optimality Condition) *Given a convex set M and a convex differentiable function $f : M \to \mathbb{R}$, $x^* \in \arg\min_x f(x)$ s.t. $x \in X$ if and only if*

$$\nabla f(x^*)^\top (y - x^*) \geq 0, \textit{ for all } y \in M.$$

Before we prove the proposition, we illustrate the geometry of the first-order optimality condition in Fig. 12.3. We use the constrained Lasso as an example. We can see that the first-order optimality condition essentially implies that the constraint M is on the one side of the hyperplane $p(y) = \nabla f(x^*)^\top (y - x^*)$. Specifically, $\nabla f(x^*)^\top (y - x^*) \geq 0$ implies that the angle between the gradient $\nabla f(x^*)$ and $y - x^*$ is always acute.

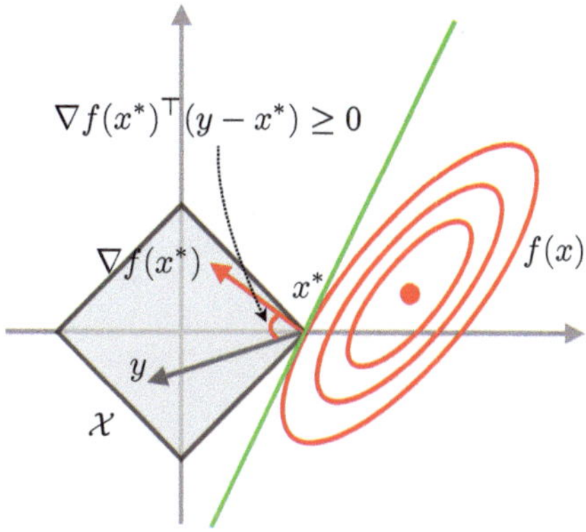

Fig. 12.3 The geometry of the first-order optimality condition $\nabla f(x^*)^\top (y - x^*) \geq 0$, for all $y \in M$ for the problem $\min_x f(x)$ s.t. $x \in M$

The optimality conditions for the convex optimization $x^* \in \arg\min_x f(x)$ s.t. $x \in M$: for any $y \in M$ (when the problem is unconstrained $M = \mathbb{R}^d$), we have

	Unconstrained	Constrained
Zero order	$f(x^*) \leq f(y)$	$f(x^*) \leq f(y)$
First order	$0 \in \partial f(x^*)$	$\nabla f(x^*)^\top (y - x^*) \geq 0$

Proof We first prove the "if" part. By the definition of subgradient, we have

$$f(y) - f(x^*) \geq \nabla f(x^*)^\top (y - x^*) \geq 0.$$

Therefore, we have $f(y) \geq f(x^*)$ and thus x^* is minimizer.

To prove the "only if" part, it suffices to show that if there exists a $y \in M$ such that $\nabla f(x^*)^\top (x^* - y) > 0$, then x^* is not a minimum. Define $h(t) = f(x^* + t(y - x^*))$. We have $h'(0) = \nabla f(x^*)^\top (y - x^*) < 0$

$$h'(0) = \nabla f(x^*)^\top (y - x^*) < 0.$$

So $h(t)$ is decreasing at 0, i.e., $f(x)$ will decrease along the direction of $y - x^*$. Therefore, x^* is not a minimizer. □

Chapter 13
Gradient Descent

13.1 Gradient Descent

In this chapter, we will start designing algorithms to solve the convex optimization

$$\min_{x \in \mathcal{X}} f(x), \text{ where } f \text{ and } M \text{ are convex.}$$

Our goal is to find the minimizer $x^* = \arg\min_{x \in M} f(x)$. Let us start with the unconstrained problem first with $M = \mathbb{R}^d$. If we start our search for x^* at some value x_0, we aim to move to the next point such that the value of $f(x)$ becomes smaller. If we go along the direction d with an infinitesimal step-size τ, the decrease of the objective function is

$$\lim_{\tau \to 0} \frac{f(x_0 + \tau d) - f(x_0)}{\tau} = \nabla f(x_0)^\top d. \tag{13.1}$$

We hope to choose the direction d such that the decrease is maximized, and from the result above, we can see that the optimal direction is $d = -\nabla f(x_0)$. Therefore, if we can only utilize the local information of $f(x)$ at the point x_0, the negative gradient $-\nabla f(x_0)$ is the steepest descent direction. This simple strategy motivates the gradient descent algorithm.

The **gradient descent algorithm** is an iterative algorithm starting at the point $x_0 \in \mathbb{R}^d$ and it iterates as

$$x_{t+1} = x_t - \eta_t \nabla f(x_t),$$

where η_t is the step-size (also called learning rate).

J. Lu, *Big Data Analysis*, https://doi.org/10.1007/978-3-032-03161-7_13

In order to guarantee good convergence properties of the gradient descent, the landscape of $f(x)$ cannot be too steep or too rough. We formalize this intuition by the following definition of L-smooth.

Definition 13.1 (L-Smooth) A continuously differentiable function f is L-smooth if

$$f(y) \leq f(x) + \nabla f(x)^\top (y - x) + \frac{L}{2} \|x - y\|_2^2,$$

for any x and y.

We can see that $f(x)+\nabla f(x)^\top(y-x)$ is the first-order Taylor expansion of $f(y)$. So the smoothness of f implies that the difference between the function value and its first-order Taylor can be controlled by the quadratic function $\frac{L}{2}\|y - x\|_2^2$. If f has the second derivative, the second-order Taylor expansion implies

$$f(y) \approx f(x) + \nabla f(x)^\top (y - x) + \frac{1}{2}(y - x)^\top \nabla^2 f(x)(y - x). \tag{13.2}$$

Comparing it with Definition 13.1, the L-smooth implies that the Hessian $\nabla^2 f(x)$ cannot be too large. In fact, the L-smooth is equivalent to the following properties:

- If f has the second derivatives, (13.2) is equivalent to $\nabla^2 f(x) \preceq L\mathbf{I}_d$ for any x, which prevents f from being too curved.
- (13.2) is equivalent to $\|\nabla f(x) - \nabla f(y)\|_2 \leq L\|x - y\|_2$, i.e., the gradient of f is Lipschitz. So L-smoothness implies that the gradient of f does not change too drastically.

For example, the least squares loss $f(\beta) = \frac{1}{2}\|Y - \mathbb{X}\beta\|_2^2$ has the Hessian matrix $\nabla^2 f(x) = \mathbb{X}\mathbb{X}^\top$. So the least squares loss $\frac{1}{2}\|Y - \mathbb{X}\beta\|_2^2$ is L-smooth if the maximum eigenvalue of $\mathbb{X}\mathbb{X}^\top$ is smaller than L. On the other hand, the ℓ_1-norm $\|x\|_1$ is not smooth as its derivative is not continuous (not mentioning L-Lipschitz) at $x = 0$.

The following theorem shows how fast the gradient descent converges.

Theorem 13.1 (Convergence of Gradient Descent) *Let f be convex and L-smooth. If we choose $\eta_t = 1/L$, then the gradient descent has*

$$f(x_t) - f(x^*) \leq \frac{2L\|x_0 - x^*\|_2^2}{t}.$$

We say the convergence rate of gradient is $O(1/t)$, or to rephrase, the gradient descent can achieve the ϵ-accuracy $f(x_t) - f(x^*) \leq \varepsilon$ within $O(1/\epsilon)$ steps. We omit the proof in the class, but we will list the proof here.

Proof By the smoothness of $f(x)$ and the gradient descent iteration $x_{t+1} = x_t - \eta_t \nabla f(x_t)$, we have

$$\begin{aligned} f(x_{t+1}) - f(x_t) &\le \nabla f(x_t)^\top (x_{t+1} - x_t) + \frac{L}{2}\|x_{t+1} - x_t\|_2^2 \\ &= -\eta_t \|\nabla f(x_t)\|_2 + \frac{\eta_t^2 L}{2}\|\nabla f(x_t)\|_2^2. \end{aligned}$$

In order to maximize the descent, we choose $\eta_t = 1/L$ to maximize the right-hand side of the inequality and get

$$f(x_{t+1}) - f(x_t) \le -\frac{1}{2L}\|\nabla f(x_t)\|_2^2. \tag{13.3}$$

This inequality shows that the objective value of the gradient descent is monotonic decreasing.

In fact, not only the objective value is decreasing, we can also show that the distance between x_t and x^* is also decreasing. To prove this, we need the following lemma derived from the L-smoothness. □

Lemma 13.1 *f is L-smooth. Then for any $x, y \in \mathbb{R}^n$, one has*

$$f(x) - f(y) \le \nabla f(x)^\top (x - y) - \frac{1}{2L}\|\nabla f(x) - \nabla f(y)\|_2^2.$$

Proof Let $z = y - \frac{1}{L}(\nabla f(y) - \nabla f(x))$. Then one has

$$\begin{aligned} f(x) - f(y) = f(x) - f(z) + f(z) - f(y) &\le \nabla f(x)^\top (x - z) \\ &\quad + \nabla f(y)^\top (z - y) + \frac{L}{2}\|z - y\|_2^2 \\ &= \nabla f(x)^\top (x - y) + (\nabla f(x) - \nabla f(y))^\top (y - z) \\ &\quad + \frac{1}{2L}\|\nabla f(x) - \nabla f(y)\|_2^2 \\ &= \nabla f(x)^\top (x - y) - \frac{1}{2L}\|\nabla f(x) - \nabla f(y)\|_2^2. \end{aligned}$$

□

Now we show that $\|x_t - x^*\|_2$ is decreasing with t. Using Lemma 13.1 one immediately gets

$$(\nabla f(x) - \nabla f(y))^\top (x - y) \ge \frac{1}{2L}\|\nabla f(x) - \nabla f(y)\|_2^2.$$

We use this inequality by setting $x = x_{t+1}, y = x^*$ as follows (together with $\nabla f(x^*) = 0$)):

$$\begin{aligned}\left\|x_{t+1}-x^*\right\|_2^2 &= \left\|x_t-\frac{1}{L}\nabla f\left(x_t\right)-x^*\right\|_2^2\\ &= \left\|x_t-x^*\right\|_2^2-\frac{2}{L}\nabla f\left(x_t\right)^\top\left(x_t-x^*\right)+\frac{1}{L^2}\left\|\nabla f\left(x_t\right)\right\|_2^2\\ &\le \left\|x_t-x^*\right\|_2^2.\end{aligned}$$

Denoting $\delta_t = f\left(x_t\right) - f\left(x^*\right)$, we can rewrite (13.3) as

$$\delta_{t+1} \le \delta_t - \frac{1}{2L}\left\|\nabla f\left(x_t\right)\right\|_2^2.$$

One also has by convexity (Definition 12.3),

$$\delta_t \le \nabla f\left(x_t\right)^\top\left(x_t-x^*\right) \le \left\|x_t-x^*\right\|_2\left\|\nabla f\left(x_t\right)\right\|_2.$$

As $\|x_t - x^*\|_2$ is decreasing with t, which with the two above displays will imply

$$\delta_{t+1} \le \delta_t - \frac{1}{2L\left\|x_0-x^*\right\|_2^2}\delta_t^2.$$

Let us see how to use this last inequality to conclude the proof. Let $\omega = \frac{1}{2L\|x_1-x^*\|_2^2}$, then

$$\omega\delta_t^2+\delta_{t+1}\le\delta_t \Leftrightarrow \omega\frac{\delta_t}{\delta_{t+1}}+\frac{1}{\delta_t}\le\frac{1}{\delta_{t+1}} \Rightarrow \frac{1}{\delta_{t+1}}-\frac{1}{\delta_t}\ge\omega \Rightarrow \frac{1}{\delta_t}\ge\omega(t-1).$$

13.2 Frank-Wolfe Algorithm

We now consider the constrained problem $\min_{x\in M} f(x)$. The gradient descent algorithm might make x_t escape the constraint $\mathcal{X}$. As illustrated in Fig. 13.1, we still want to find the steepest descent d, but we also need to make sure that x_t belongs to M for all $t = 0, 1, 2, \ldots$. Therefore, similar to (13.1), starting with $x_0 \in M$, we want to find the steepest descent $d = x - x_0$ with $x \in M$ via the following steepest descent:

$$\arg\min_{x\in\mathcal{X}}\langle\nabla f(x_0), x-x_0\rangle = \arg\min_{x\in\mathcal{X}}\langle\nabla f(x_0), x\rangle.$$

This motivates the following Frank-Wolfe algorithm.

Fig. 13.1 The illustration of the Frank-Wolfe algorithm

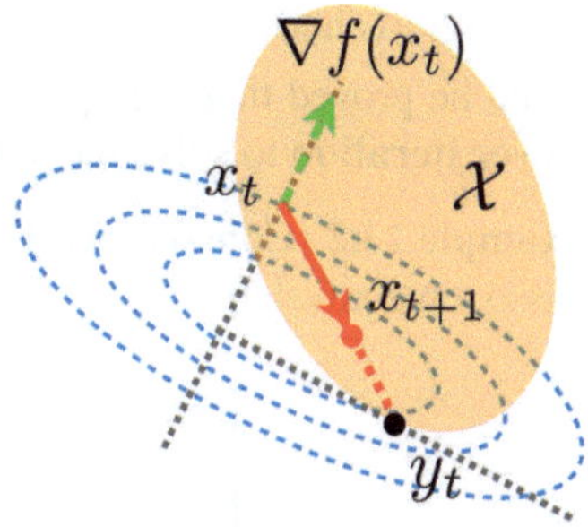

The **Frank-Wolfe algorithm** performs the update

$$y_t = \arg\min_{x \in \mathcal{X}} \langle \nabla f(x_t), x \rangle;$$

$$x_{t+1} = x_t + \eta_t (y_t - x_t),$$

where η_t is the step-size.

Example 13.1 (Power Iteration) We know that the leading eigenvector of a positive semi-definite matrix A is the solution of the following optimization problem (see Theorem 3.2):

$$\max_{\|x\|_2 \le 1} x^\top A x = \min_{\|x\|_2 \le 1} -x^\top A x.$$

Notice that $f(x) = -x^\top A x$ is concave as $A \succeq 0$. So the above problem is not convex. However, we can still use the Frank-Wolfe algorithm to solve it and find the leading eigenvector of A. Implementing the Frank-Wolfe algorithm, we need to solve

$$y_t = \arg\min_{\|x\|_2 \le 1} \langle \nabla f(x_t), x \rangle = -\frac{\nabla f(x_t)}{\|\nabla f(x_t)\|_2} = \frac{A x_t}{\|A x_t\|_2},$$

where we use the variational form of ℓ_2-norm $\arg\max_{\|x\|_2 \le 1} \langle y, x \rangle = y / \|y\|_2$. This gives us the updating rule

$$x_{t+1} = x_t + \eta_t \Big(\frac{A x_t}{\|A x_t\|_2} - x_t \Big) = (1 - \eta_t) x_t + \eta_t \frac{A x_t}{\|A x_t\|_2}.$$

We want to choose the step-size $\eta_t \in [0, 1]$ to minimize $f(x_{t+1}) = -x_{t+1}^\top Ax_{t+1}$. It can be proved that the optimal $\eta_t = 1$ and thus $x_{t+1} = \frac{Ax_t}{\|Ax_t\|_2}$, which is called **power iteration** to solve the leading eigenvector of A.

Example 13.2 (Constrained Lasso) We can see that the following two constrained Lasso are equivalent:

$$\begin{cases} \min_\beta \|Y - \mathbb{X}\beta\|_2^2 \\ \text{s.t. } \|\beta\|_1 \le \lambda, \end{cases} \Longleftrightarrow \begin{cases} \min_\beta \|Y/\lambda - \mathbb{X}\beta\|_2^2 \\ \text{s.t. } \|\beta\|_1 \le 1. \end{cases}$$

Therefore, without loss of generality, we can assume $\lambda = 1$ in the constrained Lasso. To solve it by Frank-Wolfe algorithm, we have

$$y_t = \arg\min_{\|\beta\|_1 \le 1} \langle \nabla f(\beta_t), \beta \rangle,$$

where $\nabla f(\beta_t) = \nabla_\beta \|Y - \mathbb{X}\beta_t\|_2^2 = -2\mathbb{X}^\top (Y - \mathbb{X}\beta_t)$. By the variational form of ℓ_1-norm (see Proposition 8.1), we know the dual norm of ℓ_1-norm is $\|x\|_\infty = \max_{\|y\|_1 \le 1} \langle y, x\rangle$ and

$$\arg\max_{\|y\|_1 \le 1} \langle y, x \rangle = (0, \ldots, 0, \text{sign}\,(x_{j^*}), 0, \ldots, 0)^\top, \text{ where } j^* = \arg\max_{1 \le j \le d} |x_j|.$$

Plugging this back into FW algorithm, we get

$$(y_t)_j = \begin{cases} -\text{sign}\,(\nabla_j f(\beta_t)), & \text{if } j = \arg\max_k \nabla_k f(\beta_t); \\ 0, & \text{otherwise.} \end{cases}$$

$$\beta_{t+1} = (1 - \eta_t)\beta_t + \eta_t y_t.$$

Notice that only one entry of y_t is nonzero; therefore, in each iteration, β_{t+1} at most has one more nonzero entry than x_t. So if we start at $\beta_0 = 0$, we have $\|x_t\|_0 \le t$. This implies that the algorithm is very fast and also provides further intuition as to why the ℓ_1 constraint implies sparsity in the optimal Lasso solution.

The following theorem shows how to choose the step-size in Frank-Wolfe and how fast it converges.

Theorem 13.2 (Convergence Rate of Frank-Wolfe Algorithm) *Let f be convex and L-smooth. If we choose $\eta_t = \frac{2}{t+2}$, then the Frank-Wolfe algorithm has*

$$f(x_t) - f(x^*) \le \frac{2Ld_{\mathcal{X}}^2}{t+2},$$

where $d_{\mathcal{X}}^2 = \sup_{x,y \in \mathcal{X}} \|x - y\|_2^2$.

Similar to the convergence rate of gradient descent in Theorem 13.1, for convex and smooth function, the Frank-Wolfe algorithm has the convergence rate $O(1/t)$, or to rephrase, we can achieve the ϵ-accuracy within $O(1/\epsilon)$ steps. However, unlike the constant step-size for the gradient descent, the suggested step-size of the Frank-Wolfe algorithm is shrinking and is irrelevant to L.

Proof By smoothness, we have

$$
\begin{aligned}
& f(x_{t+1}) - f(x_t) \\
& \quad \leq \nabla f(x_t)^\top (x_{t+1} - x_t) + \frac{L}{2}\|x_{t+1} - x_t\|_2^2 \\
& \quad \leq \eta_t \nabla f(x_t)^\top (y_t - x_t) + \frac{L}{2}\eta_t^2 \|y_t - x_t\|_2^2 && (\text{By } x_{t+1} - x_t = \eta_t(y_t - x_t)) \\
& \quad \leq \eta_t \nabla f(x_t)^\top (x^* - x_t) + \frac{L}{2}\eta_t^2 d_{\mathcal{X}}^2 && (\text{By } y_t = \arg\min_{x\in\mathcal{X}} \nabla f(x_t)^\top x) \\
& \quad \leq \eta_t (f(x^*) - f(x_t)) + \frac{L}{2}\eta_t^2 d_{\mathcal{X}}^2 && (\text{By convexity, Definition 10.3}).
\end{aligned}
$$

Now letting $\Delta_t = f(x_t) - f(x^*)$, we get

$$
\Delta_{t+1} - \Delta_t \leq -\eta_t \Delta_t + \frac{L}{2}\eta_t^2 d_{\mathcal{X}}^2 \implies \Delta_{t+1} \leq (1 - \eta_t)\Delta_t + \frac{L}{2}\eta_t^2 d_{\mathcal{X}}^2.
$$

We set $\eta_t = 2/(t+2)$, and it follows from induction that $\Delta_t \leq \frac{2L}{t+2} d_{\mathcal{X}}^2$. □

13.3 Accelerated Gradient Descent

From Fig. 13.2a, we notice that the trajectory of gradient descent is zigzag. This is because the steep descent direction in (13.1) only depends on the local information of the objective function and thus is short-sighted. Can we find an algorithm that converges faster than the gradient descent? We may consider the following two strategies to remedy the problems of the gradient descent we mentioned above:

- Exploit the history of the trajectory.
- Add buffers to have a smoother trajectory.

In Fig. 13.2b, instead of using the local gradient at x_t, we can use the information in the history and combine $-\nabla f(x_t)$ with the previous points $x_t - x_{t-1}$. We can see that the combined direction is more stable and less zigzag. Nesterov proposed the following accelerated gradient descent which by its name converges faster than the gradient descent.

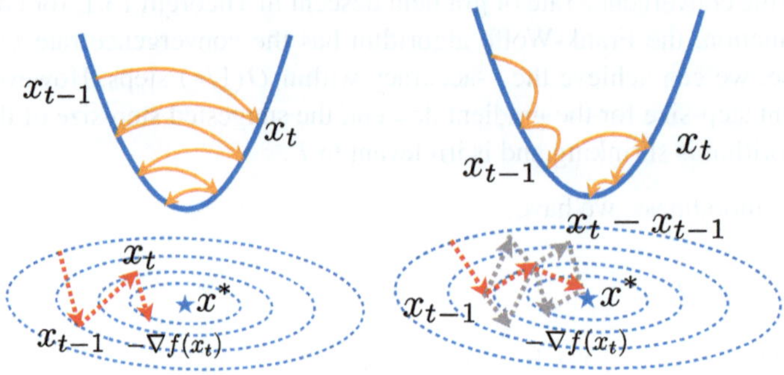

(a) (b)

Fig. 13.2 The trajectories of the gradient descent and the accelerated gradient descent. (**a**) Gradient descent. (**b**) Accelerated Gradient descent

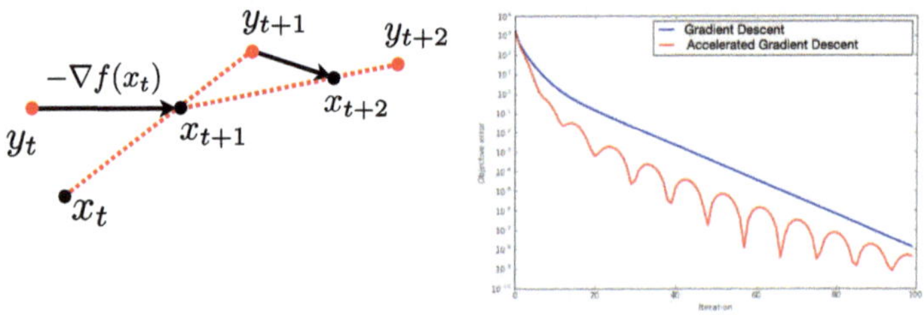

(a) (b)

Fig. 13.3 The objective values of the gradient descent are monotonically decreasing, while the AGD does not. (**a**) The trajectory of AGD. (**b**) Comparison between GD and AGD

> The **Nesterov's accelerated gradient descent** (AGD) algorithm: Initialize $x_0 = y_0$,
>
> $$x_{t+1} = y_t - \eta_t \nabla f(y_t)$$
>
> $$y_{t+1} = x_{t+1} + \frac{\lambda_t - 1}{\lambda_{t+1}}(x_{t+1} - x_t),$$
>
> where $\lambda_0 = 1$, $\lambda_t = \frac{1+\sqrt{1+4\lambda_{t-1}^2}}{2}$.

From Fig. 13.3a, we can see that compared to the gradient descent, the updating rule of the accelerated gradient descent utilizes the history of the trajectory through the $(x_{t+1} - x_t)$ term. However, the weight $\frac{\lambda_t - 1}{\lambda_{t+1}}$ is somehow mysterious. We will try to provide some explanation in the next chapter. Since the AGD utilizes the history, unlike the gradient descent, the trajectory of AGD is not monotonically decreasing towards the minimum. See Fig. 13.3b.

The following theorem shows the convergence rate of AGD.

Theorem 13.3 (Convergence of AGD) *Let f be convex and L-smooth. If we choose $\eta_t = 1/L$, then Nestorov's accelerated gradient descent has*

$$f(x_t) - f(x^*) \le \frac{2L\|x_0 - x^*\|_2^2}{(t+1)^2}$$

This means that the convergence rate of AGD is $O(1/t^2)$, which is faster than the $O(1/t)$ rate of the gradient descent in Theorem 13.1. Or to rephrase, for convex and smooth objective functions, we can achieve ϵ-accuracy within $O(1/\sqrt{\epsilon})$ steps.

We will show the proof of this theorem and give more insights of the AGD later.

Chapter 14
Proximal Gradient Descent

14.1 Proximal Perspective

In the previous chapter, we introduce the gradient descent and accelerated gradient algorithm to solve the unconstrained optimization. We show the convergence rates of these two algorithms when the objective function is smooth. However, in Lasso $\min_\beta \frac{1}{2}\|Y-\mathbb{X}\beta\|_2^2+\lambda\|\beta\|_1$, the ℓ_1-norm penalty term is not smooth. For many other high-dimensional M-estimators, e.g., these examples we introduced in Chap. 9, they can be formulated as the optimization problem:

$$\min_{x \in \mathbb{R}^d} F(x) = \min_{x \in \mathbb{R}^d} f(x) + h(x),$$

where $f(x)$ is the loss function, which is typically convex and smooth, and $h(x)$ is the penalty term, which is convex but usually non-differentiable. If we directly apply the (sub-)gradient descent algorithm to these problems, the convergence rate will be deteriorated by the non-smooth part of the objective function.

In this chapter, we will focus on the algorithms solving this type of composite loss and expect them to converge as fast as the gradient descent for smooth objective functions. Before we introduce the new algorithm, we first start with giving some new insight of the gradient descent

$$x_{t+1} = x_t - \eta_t \nabla f(x_t).$$

In the previous chapter, we motivated the gradient descent by showing the $-\nabla f(x_t)$ is the steepest descent direction. An alternative perspective to understand the gradient descent is that we aim to approximate the objective function $f(x)$ around $x = x_t$ by a quadratic function

J. Lu, *Big Data Analysis*, https://doi.org/10.1007/978-3-032-03161-7_14

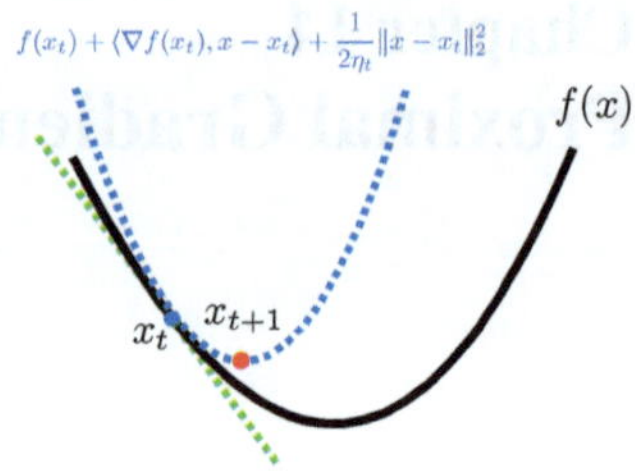

Fig. 14.1 The proximal perspective of gradient descent

$$f(x) \approx \underbrace{f(x_t) + \langle \nabla f(x_t), x - x_t \rangle}_{\text{1st order Taylor expansion}} + \underbrace{\frac{1}{2\eta_t}\|x - x_t\|_2^2}_{\text{proximal term}}.$$

Instead of minimizing $f(x)$, we first minimize its quadratic approximation

$$x_{t+1} = \arg\min_{x \in \mathbb{R}^d} \left\{ f(x_t) + \langle \nabla f(x_t), x - x_t \rangle + \frac{1}{2\eta_t}\|x - x_t\|_2^2 \right\}; \tag{14.1}$$

see Fig. 14.1 for an illustration.

The problem in (14.1) has a closed form solution which is exactly gradient descent:

$$\begin{aligned} x_{t+1} &= \arg\min_{x \in \mathbb{R}^d} \left\{ f(x_t) + \langle \nabla f(x_t), x - x_t \rangle + \frac{1}{2\eta_t}\|x - x_t\|_2^2 \right\} \\ &= \arg\min_{x \in \mathbb{R}^d} \left\{ \frac{1}{2}\|x - (x_t - \eta_t \nabla f(x_t))\|_2^2 \right\} \\ &= x_t - \eta_t \nabla f(x_t). \end{aligned}$$

So from the proximal perspective, in each iteration, the gradient descent tries to minimize a local quadratic approximation of the objective function.

Now let us go back to the composite loss $F(x) = f(x) + h(x)$. We can modify the proximal perspective of the gradient descent in (14.1) as

$$\begin{aligned} x_{t+1} &= \arg\min_{x \in \mathbb{R}^d} \left\{ f(x_t) + \langle \nabla f(x_t), x - x_t \rangle + \frac{1}{2\eta_t}\|x - x_t\|_2^2 + h(x) \right\} \\ &= \arg\min_{x \in \mathbb{R}^d} \left\{ \frac{1}{2}\|x - (x_t - \eta_t \nabla f(x_t))\|_2^2 + \eta_t h(x) \right\}. \end{aligned}$$

We then will have the following algorithm to solve $\min_{x \in \mathbb{R}^d} f(x) + h(x)$.

Define the **proximal operator** as

$$\mathrm{prox}_h(x) = \arg\min_{z\in\mathbb{R}^d} \left\{ \frac{1}{2}\|z - x\|_2^2 + h(z) \right\}.$$

The **proximal gradient descent** can be written as

$$x_{t+1} = \mathrm{prox}_{\eta_t h}\big(x_t - \eta_t \nabla f(x_t)\big).$$

We will give a few examples of the proximal gradient descent.

Example 14.1 (Constrained Optimization) Although the proximal gradient descent is designed for an unconstrained problem, we can reformulate the constrained optimization $\min_{x\in M} f(x)$ as the unconstrained composite form $\min_{x\in\mathbb{R}^d} f(x) + h(x)$, where the indicator function

$$h(x) = \begin{cases} 0, & \text{if } x \in \mathcal{X}; \\ \infty, & \text{if } x \notin \mathcal{X}. \end{cases}$$

We can solve the proximal operator

$$\mathrm{prox}_h(x) = \arg\min_{z\in\mathbb{R}^d} \frac{1}{2}\|z - x\|_2^2 + h(z) = \arg\min_{z\in\mathcal{X}} \|z - x\|_2^2,$$

which is projecting x to the constraint M. Thus, the proximal gradient descent gives us the projected gradient descent algorithm to solve the constrained optimization.

The **projected gradient descent** is

$$y_t = x_t - \eta_t \nabla f(x_t);$$

$$x_{t+1} = \arg\min_{x\in\mathcal{X}} \|y_t - x\|_2^2.$$

We illustrated the projected gradient descent in Fig. 14.2. Now we have two algorithms to solve the constrained optimization, the Frank-Wolfe algorithm and the projected gradient descent. Both algorithms need to solve a sub-problem in the iterations. The projected gradient descent need to find the projection, while the

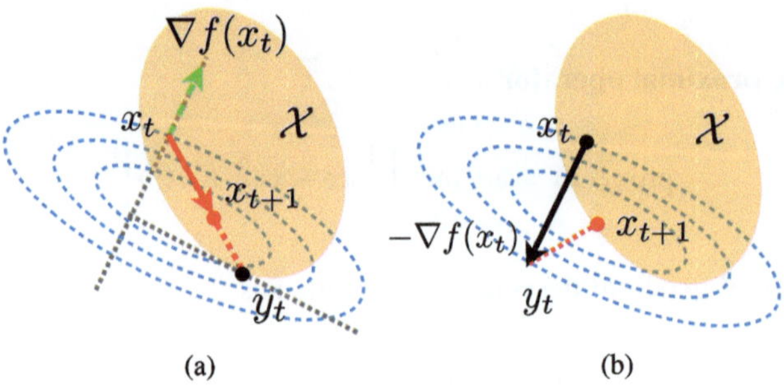

Fig. 14.2 The comparison between the projected gradient descent and the Frank-Wolfe algorithm. (**a**) Frank-Wolfe algorithm. (**b**) Projected gradient descent

Frank-Wolfe algorithm needs to solve $\arg\min_{x\in\mathcal{X}}\langle\nabla f(x_{t-1}), x\rangle$. In practice, we will tend to choose the algorithm such that the sub-problem is easier to compute.

Example 14.2 (Lasso) The ℓ_1-penalized optimization has the objective function $\min_x f(x) + \lambda\|x\|_1$. The proximal operator

$$\text{prox}_h(x) = \arg\min_{z\in\mathbb{R}^d} \frac{1}{2}\|z - x\|_2^2 + \lambda\|z\|_1$$

becomes the soft-threshold operator

$$[\text{prox}_h(x)]_j = [\text{Soft-Threshold}(x, \lambda)]_j = \begin{cases} x_j - \lambda, & \text{if } x_j \geq \lambda; \\ x_j + \lambda, & \text{if } x_j \leq -\lambda; \\ 0, & \text{otherwise,} \end{cases} \tag{14.2}$$

for all $j = 1, \ldots, d$. We visualize the proximal operator of ℓ_1-norm in Fig. 14.3. The proximal gradient descent reduces to the following algorithm.

The iterative shrinkage-thresholding algorithm (**ISTA**) solves the Lasso problem $\min_{x\in\mathbb{R}^d} f(x) + \lambda\|x\|_1$ as

$$y_t = x_t - \eta_t\nabla f(x_t);$$
$$x_{t+1} = \text{Soft-Threshold}(y_t, \lambda\eta_t).$$

Even if the objective function $F(x) = f(x) + h(x)$ has the non-smooth $h(x)$, the following theorem shows that the proximal gradient descent has the same convergence rate $O(1/t)$ as the gradient descent.

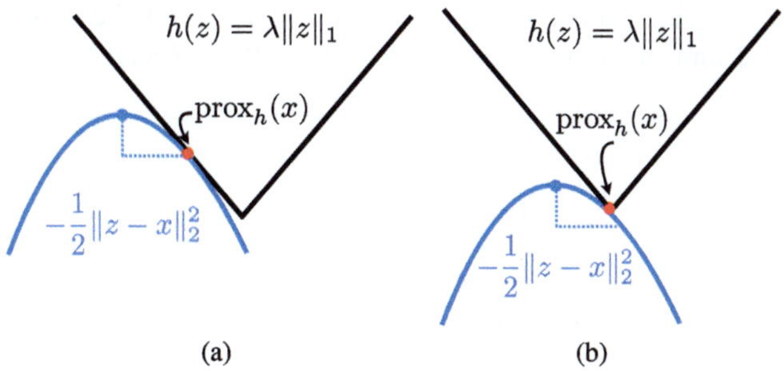

Fig. 14.3 The proximal operator of ℓ_1-norm and the soft-thresholding. (**a**) $|x| \geq \lambda$. (**b**) $|x| < \lambda$

Theorem 14.1 (Convergence Rate of Proximal Gradient Descent) *Suppose f is convex and L-smooth and h is convex. If $\eta_t = 1/L$, the proximal gradient descent has*

$$F(x_t) - F(x^*) \leq \frac{L\|x_0 - x^*\|_2^2}{2t}. \tag{14.3}$$

Before we prove the theorem, we need to establish a fundamental inequality in the following lemma.

Lemma 14.1 *Suppose f is convex and L-smooth and h is convex. For any $x, y \in \mathbb{R}^d$, let $y^+ = \text{prox}_{\frac{1}{L}h}(y - \frac{1}{L}\nabla f(y))$, then*

$$F(y^+) - F(x) \leq \frac{L}{2}\|x - y\|_2^2 - \frac{L}{2}\|x - y^+\|_2^2.$$

This lemma provides a very useful inequality which bounds the objective function by the distance of points.

Proof of Lemma 14.1 We have

$$\begin{aligned}
F(y^+) - F(x) &= \underbrace{f(y^+) - f(y)}_{L\text{-smooth}} + \underbrace{f(y) - f(x)}_{\text{convexity}} + \underbrace{h(y^+) - h(x)}_{\text{convexity}} \\
&\leq \nabla f(y)^\top (y^+ - y) + \frac{L}{2}\|y^+ - y\|_2^2 + \nabla f(y)^\top (y - x) \\
&\quad + \partial h(y^+)^\top (y^+ - x) \\
&= \langle \nabla f(y) + L(y^+ - y) + \partial h(y^+), y^+ - x \rangle \\
&\quad + \frac{L}{2}\|y^+ - y\|_2^2 - L\langle y^+ - y, y^+ - x \rangle,
\end{aligned}$$

where in the first inequality, we apply the inequalities on the L-smoothness in Definition 11.2 and the subgradient in Definition 10.3. Recall that we define y^+ as

$$\begin{aligned} y^+ &= \text{prox}_{\frac{1}{L}h}\Big(y - \frac{1}{L}\nabla f(y)\Big) \\ &= \arg\max_z \underbrace{f(y) + \langle \nabla f(y), z - y\rangle + \frac{L}{2}\|z - y\|_2^2 + h(z)}_{\phi(z)}. \end{aligned}$$

By the first optimality condition, we have

$$\partial\phi(y^+) = \nabla f(y) + L(y^+ - y) + \partial h(y^+) = 0.$$

Therefore, we have

$$F(y^+) - F(x) \le \frac{L}{2}\|y_t - y\|_2^2 - L\langle y^+ - y, y^+ - x\rangle = \frac{L}{2}\|x - y\|_2^2 - \frac{L}{2}\|x - y^+\|_2^2,$$

where in the last equality, we use the identity $\|a\|_2^2 - 2\langle a, b\rangle = \|a - b\|_2^2 - \|b\|_2^2$. □

Now we can go back to the proof of the convergence rate of the proximal gradient descent.

Proof of Theorem 14.1 First, we prove that the values of the objective function are decreasing with t. We apply Lemma 14.1 by letting $x = x_t$, $y = x_t$, $y^+ = x_{t+1}$ and have

$$F(x_{t+1}) - F(x_t) \le -\frac{L}{2}\|x_{t+1} - x_t\|_2^2 < 0.$$

So $F(x_t)$ decreases as t increases.

Second, we apply Lemma 14.1 again by letting $x = x^*$, $y = x_t$, $y^+ = x_{t+1}$ and have

$$F(x_{t+1}) - F(x^*) \le \frac{L}{2}\|x_t - x^*\|_2^2 - \frac{L}{2}\|x_{t+1} - x^*\|_2^2.$$

We sum up the above inequality from 0 to $t-1$ and get

$$\sum_{s=0}^{t-1}(F(x_{s+1}) - F(x^*)) \le \frac{L}{2}\|x_0 - x^*\|_2^2 - \frac{L}{2}\|x_t - x^*\|_2^2.$$

As in the first part, we showed $F(x_s) - F(x^*)$ decreases as t increases, we have

$$F(x_t) - F(x^*) \le \frac{L\|x_0 - x^*\|_2^2}{2t}.$$

□

14.2 Accelerated Proximal Gradient Descent

Theorem 14.1 shows that the proximal gradient descent has the convergence rate $O(1/t)$, which is same as the gradient descent. In the previous chapter, we introduced the Nesterov's accelerated gradient descent and shows that it converges faster than the gradient descent with the rate $O(1/t^2)$.

We can also apply Nesterov's idea to the proximal gradient descent and derive the following algorithm.

The **accelerated proximal gradient descent** algorithm: Initialize $x_0 = y_0$,

$$x_{t+1} = \text{prox}_{\eta_t h}\big(y_t - \eta_t \nabla f(y_t)\big);$$

$$y_{t+1} = x_{t+1} + \frac{\lambda_t - 1}{\lambda_{t+1}}(x_{t+1} - x_t),$$

where $\lambda_0 = 1$, $\lambda_t = \frac{1+\sqrt{1+4\lambda_{t-1}^2}}{2}$.

For example, we can accelerate ISTA for Lasso $\min_x f(x) + \lambda\|x\|_1$ in Example 14.2 via the following algorithm.

The fast iterative shrinkage-thresholding algorithm (**FISTA**) solves the Lasso problem $\min_{x\in\mathbb{R}^d} f(x) + \lambda\|x\|_1$ as

$$x_{t+1} = \text{Soft-Threshold}(y_t - \eta_t \nabla f(y_t), \lambda\eta_t);$$

$$y_{t+1} = x_{t+1} + \frac{\lambda_t - 1}{\lambda_{t+1}}(x_{t+1} - x_t),$$

where $\lambda_0 = 1$, $\lambda_t = \frac{1+\sqrt{1+4\lambda_{t-1}^2}}{2}$ and $x_0 = y_0$.

We can show that the convergence rate of the accelerated proximal gradient descent is $O(1/t^2)$.

Theorem 14.2 (Convergence Rate of Accelerated Proximal Gradient Descent) *Suppose f is convex and L-smooth and h is convex. If $\eta_t = 1/L$, the accelerated proximal gradient descent has*

$$F(x_t) - F(x^*) \le \frac{2L\|x_0 - x^*\|_2^2}{(t+1)^2}.$$

When $h = 0$, the accelerated proximal gradient descent reduces to AGD. So Theorem 13.3 is a special case of Theorem 14.2.

Proof Recall that unlike the gradient descent, the AGD is not monotonically decreasing. So our idea to prove the theorem is to construct a Lyapunov function which is monotonically decreasing using the accelerated proximal gradient descent.

The following lemma gives the construction of the Lyapunov function.

Lemma 14.2 (Lyapunov Function) *Let* $u_t = \lambda_{t-1}x_t - (x^* + (\lambda_{t-1} - 1)x_{t-1})$. *We define the Lyapunov function*

$$L_t = \|u_t\|_2^2 + \frac{2}{L}\lambda_{t-1}^2(F(x_t) - F(x^*)). \tag{14.4}$$

Then L_t *is decreasing with* t, *i.e.,* $L_t \le L_{t-1} \le \ldots \le L_1$.

We first focus on the proof of Theorem 14.2 and leave the proof of Lemma 14.2 to the end of this chapter.

We apply Lemma 14.1 by letting $x = x^*$, $y = y_{t-1}$, $y^+ = x_t$ and have

$$F(x_t) - F(x^*) \le \frac{L}{2}\left(\|y_{t-1} - x^*\|_2^2 - \|x_t - x^*\|_2^2\right).$$

Let $t = 1$, and we have

$$\frac{2}{L}(F(x_1) - F(x^*)) \le \|y_0 - x^*\|_2^2 - \|x_1 - x^*\|_2^2 = \|x_0 - x^*\|_2^2 - \|x_1 - x^*\|_2^2,$$

as we initialize $x_0 = y_0$. Recall the definition of u_t and L_t in Lemma 14.2. As $u_1 = \lambda_0 x_1 - (x^* + (\lambda_0 - 1)x_0) = x_1 - x^*$ and $\lambda_0 = 1$, we have

$$\begin{aligned} L_1 &= \|u_1\|_2^2 + \frac{2\lambda_0^2}{L}(F(x_1) - F(x^*)) = \|x_1 - x^*\|_2^2 + \frac{2}{L}(F(x_1) - F(x^*)) \\ &\le \|x_0 - x^*\|_2^2. \end{aligned}$$

By Lemma 14.2, the Lyapunov function is decreasing $L_t \le L_1$, i.e.,

$$L_t = \|u_t\|_2^2 + \frac{2}{L}\lambda_{t-1}^2(F(x_t) - F(x^*)) \le L_1 \le \|x_0 - x^*\|_2^2.$$

This implies that

$$F(x_t) - F(x^*) \le \frac{L\|x_0 - x^*\|_2^2}{2\lambda_{t-1}^2}.$$

By induction, we can show that $\lambda_t \geq \frac{t+2}{2}$, and therefore we have the convergence rate

$$F(x_t) - F(x^*) \leq \frac{2L\|x_0 - x^*\|_2^2}{(t+1)^2}.$$

□

We complete our chapter by showing that the Lyapunov function is monotonically decreasing.

Proof of Lemma 14.2 We apply Lemma 14.1 by letting $x = \lambda_t{}^{-1}x^* + (1-\lambda_t{}^{-1})x_t$, $y^+ = y_t$, $y_t = x_{t+1}$ and have

$$\begin{aligned} &F(x_{t+1}) - F\big(\lambda_t{}^{-1}x^* + (1-\lambda_t{}^{-1})x_t\big) \\ &\leq \tfrac{L}{2}\|\lambda_t{}^{-1}x^* + (1-\lambda_t{}^{-1})x_t - y_t\|_2^2 - \tfrac{L}{2}\|\lambda_t{}^{-1}x^* + (1-\lambda_t{}^{-1})x_t - x_{t+1}\|_2^2 \\ &= \tfrac{L}{2\lambda_t^2}(\|x^* + (\lambda_t - 1)x_t - \lambda_t y_t\|_2^2 - \|x^* + (\lambda_t - 1)x_t - \lambda_t x_{t+1}\|_2^2). \end{aligned} \tag{14.5}$$

Recall in Lemma 14.2 we define $u_{t+1} = \lambda_t x_{t+1} - (x^* + (\lambda_t - 1)x_t)$ and the AGD gives $y_t = x_t + \frac{\lambda_{t-1}-1}{\lambda_t}(x_t - x_{t-1})$. So

$$\begin{aligned} \lambda_t y_t - (x^* + (\lambda_t - 1)x_t) &= \lambda_t x_t + (\lambda_{t-1} - 1)(x_t - x_{t-1}) - (x^* + (\lambda_t - 1)x_t) \\ &= \lambda_{t-1}x_t - (x^* + (\lambda_{t-1} - 1)x_{t-1}) = u_t. \end{aligned}$$

By (14.5), we have

$$\begin{aligned} &F(x_{t+1}) - F\big(\lambda_t{}^{-1}x^* + (1-\lambda_t{}^{-1})x_t\big) \\ &\leq \frac{L}{2\lambda_t^2}(\|x^* + (\lambda_t - 1)x_t - \lambda_t y_t\|_2^2 - \|x^* + (\lambda_t - 1)x_t - \lambda_t x_{t+1}\|_2^2) \\ &= \frac{L}{2\lambda_t^2}(\|u_t\|_2^2 - \|u_{t+1}\|_2^2). \end{aligned}$$

Therefore, the above inequality derives

$$\begin{aligned} \frac{L}{2}(\|u_t\|_2^2 - \|u_{t+1}\|_2^2) &\geq \lambda_t^2\Big[F(x_{t+1}) - F\big(\lambda_t{}^{-1}x^* + (1-\lambda_t{}^{-1})x_t\big)\Big] \\ &\geq \lambda_t^2\Big[F(x_{t+1}) - \lambda_t{}^{-1}F(x^*) - (1-\lambda_t{}^{-1})F(x_t)\Big] \\ &= \lambda_t^2(F(x_{t+1} - F(x^*)) - (\lambda_t^2 - \lambda_t)(F(x_t) - F(x^*)), \end{aligned}$$

where we use the convexity of $F(x)$ in the second inequality. Recall the sequence

$$\lambda_t = \frac{1+\sqrt{1+4\lambda_{t-1}^2}}{2} \implies (2\lambda_t - 1)^2 = 1 + 4\lambda_{t-1}^2 \implies \lambda_{t-1}^2 = \lambda_t^2 - \lambda_t.$$

Therefore, we have

$$\begin{aligned}\frac{L}{2}(\|u_t\|_2^2 - \|u_{t+1}\|_2^2) &\geq \lambda_t^2(F(x_{t-1}) - F(x^*)) - (\lambda_t^2 - \lambda_t)(F(x_t) - F(x^*)) \\ &= \lambda_t^2(F(x_{t-1}) - F(x^*)) - \lambda_{t-1}^2(F(x_t) - F(x^*)).\end{aligned}$$

Recall the Lyapunov function $L_t = \|u_t\|_2^2 + \frac{2}{L}\lambda_{t-1}^2(F(x_t) - F(x^*))$ and thus the above inequality implies $L_t \geq L_{t+1}$. □

Chapter 15
Mirror Descent

15.1 Bregman Divergence

In the previous chapter, we introduce the proximal perspective of the gradient descent. To minimize $f(x)$, we approximate the objective function $f(x)$ around $x = x_t$ by a quadratic function

$$f(x) \approx \underbrace{f(x_t) + \langle \nabla f(x_t), x - x_t \rangle}_{\text{1st order Taylor expansion}} + \underbrace{\frac{1}{2\eta_t}\|x - x_t\|_2^2}_{\text{proximal term}}.$$

We consider the constrained optimization $\min_{x \in M} f(x)$. If we start at x_t, we can update the next step via minimizing the quadratic approximation

$$\begin{aligned} x_{t+1} &= \arg\min_{x \in M} \left\{ f(x_t) + \langle \nabla f(x_t), x - x_t \rangle + \frac{1}{2\eta_t}\|x - x_t\|_2^2 \right\} \\ &= \arg\min_{x \in M} \left\{ \frac{1}{2}\|x - (x_t - \eta_t \nabla f(x_t))\|_2^2 \right\}. \end{aligned}$$

If there is no constrained, i.e., $M = \mathbb{R}^d$, the above step $x_{t+1} = x_t - \eta_t \nabla f(x_t)$. Otherwise, it is the projected gradient descent. We also notice that if there is no proximal term, it reduces to the Frank-Wolfe algorithm. We add the proximal term $\frac{1}{2\eta_t}\|x - x_t\|_2^2$ in the approximation to prevent x_{t+1} from being too far away from x_t. Otherwise, if there is no constraint, $\min_{x \in \mathbb{R}^d} \nabla f(x_t)^\top x = -\infty$. A natural question is why we have to use the ℓ_2-norm in the proximal term. Is it possible to use another distance?

The following simple example shows that we can probably get a better algorithm by changing the distance in the proximal term.

J. Lu, *Big Data Analysis*, https://doi.org/10.1007/978-3-032-03161-7_15

Example 15.1 Consider the quadratic optimization

$$\min_{x\in\mathbb{R}^d} f(x) = \min_{x\in\mathbb{R}^d} \frac{1}{2}(x-x^*)^\top Q(x-x^*),$$

where Q is a positive definite matrix.

If we use the ℓ_2-norm in the proximal term, we have the gradient descent

$$\begin{aligned} x_{t+1} &= \arg\min_{x\in\mathbb{R}^d}\left\{f(x_t) + \langle \nabla f(x_t), x - x_t\rangle + \frac{1}{2\eta_t}\|x - x_t\|_2^2\right\} \\ &= x_t - \eta_t Q(x_t - x^*). \end{aligned}$$

In Fig. 15.1a, we can see that the trajectory of the gradient descent is zigzag.

On the other hand, we can imagine why the gradient descent has a zigzag trajectory is because the ℓ_2-norm is not the perfect distance for the objective $f(x) = \frac{1}{2}(x-x^*)^\top Q(x-x^*)$. The contour is scaled by the matrix Q. What if we consider the norm $\|x\|_Q^2 = x^\top Qx$ in the proximal term? We therefore update x_{t+1} as

$$\begin{aligned} x_{t+1} &= \arg\min_{x\in\mathbb{R}^d}\left\{f(x_t) + \langle \nabla f(x_t), x - x_t\rangle + \frac{1}{2\eta_t}\|x - x_t\|_Q^2\right\} \\ &= \arg\min_{x\in\mathbb{R}^d}\left\{f(x_t) + \langle \nabla f(x_t), x - x_t\rangle + \frac{1}{2\eta_t}(x - x_t)^\top Q(x-x_t)\right\} \\ &= x_t - \eta_t Q^{-1}\nabla f(x_t) = x_t - \eta_t(x_t - x^*). \end{aligned}$$

In Fig. 15.1b, we can see the descent direction directly points to the minimizer x^* which gives us a much faster algorithm.

The previous example shows that we need to find a better distance fitting the geometry of the problem. This motivates us to define the following distance.

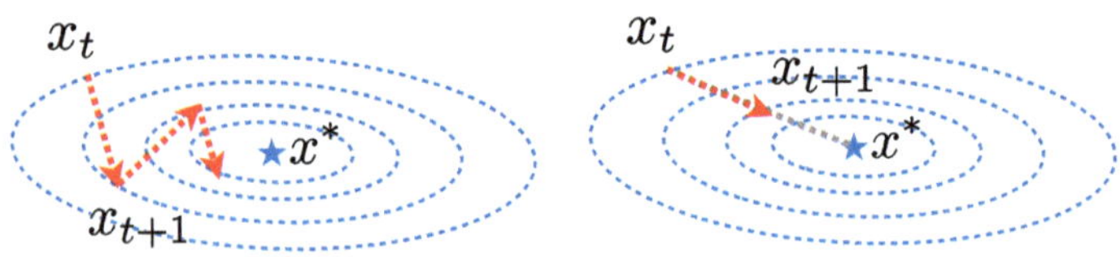

Fig. 15.1 The trajectory using the ℓ_2-norm versus using the $\|\cdot\|_Q$-norm in the proximal term. (**a**) $\|\cdot\|_2$-norm. (**b**) $\|\cdot\|_Q$-norm

Definition 15.1 (Bregman Divergence) Let $\varphi : M \to \mathbb{R}$ be a convex[1] and differentiable function, then the Bregman divergence between x and z is

$$D_\varphi(x, z) = \varphi(x) - \varphi(z) - \langle \nabla\varphi(z), x - z\rangle.$$

By convexity, $D_\varphi(x, z) \geq 0$, so it is a kind of distance. Moreover, by mean-value theorem, $\varphi(x)-\varphi(z)-\langle\nabla\varphi(z), x-z\rangle \approx \frac{1}{2}(x-z)^\top\nabla^2\varphi(\xi)(x-z)$, where ξ depends on x and z. So the Bregman divergence is like a generalization of the quadratic norm $\|\cdot\|_Q$ in Example 15.1. In fact, when $\varphi(x) = \frac{1}{2}x^\top Qx$, we can easily see that $D_\varphi(x, z) = (x - z)^\top Q(x - z)$.

15.2 Mirror Descent

If we replace the ℓ_2-norm in the proximal term by the Bregman divergence, we have

$$x_{t+1} = \arg\min_{x\in M}\left\{f(x_t) + \langle\nabla f(x_t), x - x_t\rangle + \frac{1}{\eta_t}D_\varphi(x, x_t)\right\}.$$

This motivates the following algorithm.

The **mirror descent** algorithm is

$$x_{t+1} = \arg\min_{x\in M}\left\{\langle\nabla f(x_t), x\rangle + \frac{1}{\eta_t}D_\varphi(x, x_t)\right\}.$$

Example 15.1 shows that we need to consider the Bregman divergence other than ℓ_2-norm due to the geometry of the objective function. However, for most of the cases, we need to find a proper Bregman divergence due to the geometry of the constraint. The following example illustrates how to choose the proper Bregman divergence under a specific constraint.

Example 15.2 (Probability Simplex) In many statistical problems, we need to estimate the probability mass function of a discrete distribution. So the parameters are constrained to the the **probability simplex**, defined as

[1] Actually, a standard definition of the Bregman divergence needs to assume φ is strictly convex, i.e., $\varphi((1-\gamma)x + \gamma y) < (1-\gamma)\varphi(x) + \gamma\varphi(y)$, for any $x, y \in M, \gamma \in (0, 1)$. If φ is strictly convex, then $D_\varphi(x, z) = 0$ if and only if $z = x$.

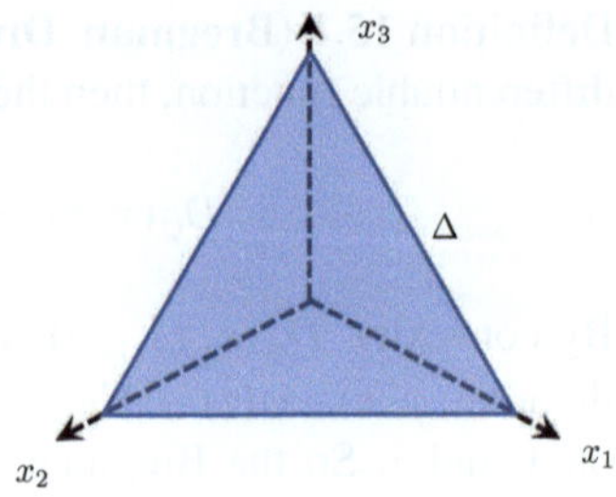

Fig. 15.2 The two-dimensional probability simplex in three-dimensional space

$$\Delta = \left\{ x \in \mathbb{R}^d \;\middle|\; \sum_{i=1}^{d} x_i = 1, x_i \geq 0 \text{ for all } i = 1, \dots, d \right\}.$$

See a visualization of a two-dimensional probability simplex in three-dimensional space in Fig. 15.2.

A widely used Bregman divergence for the probability simplex utilizes φ as the negative entropy:

$$\varphi(x) = \sum_{i=1}^{d} x_i \log x_i.$$

By definition, the corresponding Bregman divergence becomes

$$\begin{aligned} D_\varphi(x, z) &= \varphi(x) - \varphi(z) - \langle \nabla\varphi(z), x - z \rangle \\ &= \sum_{i=1}^{d} x_i \log x_i - \sum_{i=1}^{d} z_i \log z_i - \sum_{i=1}^{d} (1 + \log z_i)(x_i - z_i) \\ &= \sum_{i=1}^{d} x_i \log x_i - \sum_{i=1}^{d} (x_i - z_i) - \sum_{i=1}^{d} x_i \log z_i \\ &= \sum_{i=1}^{d} x_i \log \frac{x_i}{z_i}, \end{aligned}$$

where the last equality is due to $\sum_{i=1}^{d} x_i = \sum_{i=1}^{d} z_i = 1$. We also call $D_\varphi(x, z) = \sum_{i=1}^{d} x_i \log \frac{x_i}{z_i}$ as the Kullback-Leibler (KL) divergence, denoted as $D_{\mathrm{KL}}(x \| z)$. In fact, we have been using the KL divergence all the time while we apply the maximum log-likelihood estimator

$$\arg\max_\theta \mathbb{E}_{\theta^*}[\log P_\theta(x)] = \arg\max_\theta \mathbb{E}_{\theta^*} \log \frac{P_\theta(x)}{P_{\theta^*}(x)} = \arg\min_\theta D_{\mathrm{KL}}(P_\theta \| P_{\theta^*}).$$

Therefore, the maximum log-likelihood estimator is essentially finding a distribution P_θ closest under the KL divergence to the true distribution P_{θ^*}. From this example, we can also see that the Bregman divergence is not necessarily symmetric, i.e., $D_\varphi(x, z) \neq D_\varphi(z, x)$.

Now let us consider the constrained optimization problem $\min_{x\in\Delta} f(x)$. So far, we have learned three algorithms solving the constrained optimization:

1. **Frank-Wolfe algorithm**: We can see that the Frank-Wolfe algorithm is essentially the mirror descent with $\varphi = 0$. It involves solving a sub-problem

$$\min_{x\in\Delta}\langle \nabla f(x_t), x\rangle,$$

which is a linear programming. There is no closed form solution so we need to solve a linear programming in each iteration which is time-consuming.

2. **Projected gradient descent**: We can see that the projected gradient descent algorithm is essentially the mirror descent that uses the ℓ_2-norm as the proximal term, i.e., $\varphi(x) = \frac{1}{2}\|x\|_2^2$. It involves solving a sub-problem

$$\min_{x\in\Delta}\|x - (x_t - \eta_t \nabla f(x_t))\|_2^2,$$

which is a quadratic programming without closed form solution either.

3. **Mirror Descent**: If we use the KL divergence $D_{\mathrm{KL}}(x\|x_t)$ in the mirror descent, the sub-problem becomes

$$\arg\min_{x\in\Delta}\left\{\langle \nabla f(x_t), x\rangle + \frac{1}{\eta_t}\sum_{i=1}^{d} x_i \log\frac{x_i}{x_i^t}\right\}. \tag{15.1}$$

The sub-problem seems more complicated than the Frank-Wolfe and the projected gradient descent; however, unlike the linear or quadratic programming, (15.1) actually has a closed form solution.

We will derive the solution to (15.1) now. Apply the Lagrange multiplier

$$L(x,\lambda) = \langle \nabla f(x_t), x\rangle + \frac{1}{\eta_t}\sum_{i=1}^{d} x_i \log\frac{x_i}{x_i^t} + \lambda\Big(\sum_{i=1}^{d} x_i - 1\Big).$$

Then we have

$$\frac{\partial L(x,\lambda)}{\partial x_i}\Big|_{x=x_{t+1}} = \nabla_i f(x_t) + \frac{1}{\eta_t}\big[(1+\log x_i^{t+1}) - \log x_i^t\big] + \lambda = 0,$$

which implies that

$$x_i^{t+1} = x_i^t \exp(-\eta_t \nabla_i f(x_t)) \exp(-\eta_t\lambda - 1).$$

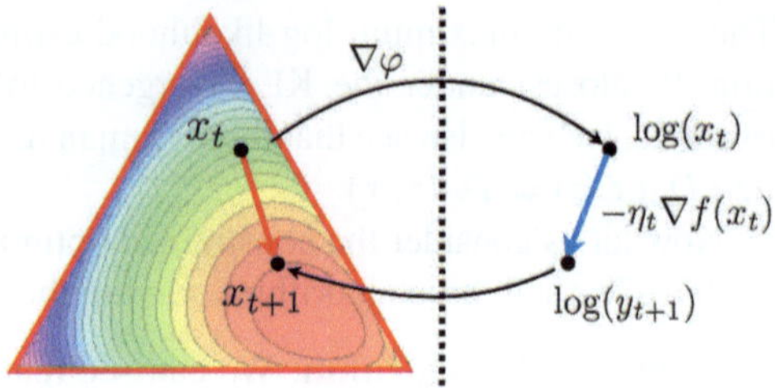

Fig. 15.3 The visualization of the mirror descent

Apply the constraint $\sum_{i=1}^{d} x_i^{t+1} = 1$ and therefore we have (Fig. 15.3)

$$x_i^{t+1} = \frac{x_i^t \exp(-\eta_t \nabla_i f(x_t))}{\sum_{j=1}^{d} x_j^t \exp(-\eta_t \nabla_j f(x_t))}, \text{ for } i = 1, \ldots, d.$$

We can see that the mirror descent has the closed form and it updates like

$$\log(y_{t+1}) = \log(x_t) - \eta_t \nabla f(x_t),$$
$$x_{t+1} = y_{t+1}/\|y_{t+1}\|_1.$$

So we can see why we call the algorithm as the mirror descent. We run the gradient descent in the "mirror space" of $\log(x_t)$ and map it back to the probability simplex via normalization.

15.3 Nesterov's Smoothing

We have introduce the proximal algorithm to solve the problem

$$\min_x f(x) + h(x).$$

When f is smooth, the accelerated proximal gradient descent has the convergence rate $O(1/t^2)$. What if the objective is not smooth? For example, in Chap. 11, we introduce the square-root Lasso

$$\min_\beta \|Y - \mathbb{X}\beta\|_2 + \lambda\|\beta\|_1.$$

We know that the ℓ_2-norm $\|x\|_2$ is not differentiable at 0. Can we somehow approximate the non-smooth function by the smooth function? In fact, Nesterov's smoothing idea is to:

1. Approximate the non-smooth objective function by a smooth function.

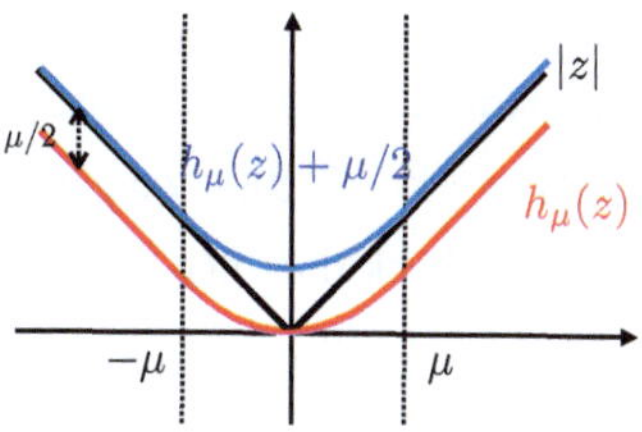

Fig. 15.4 The Huber loss function

2. Minimize the smooth approximation via (proximal) gradient descent or (proximal) accelerated gradient descent.

We rigor the idea above via the following definition.

Definition 15.2 A convex function f is (α, β)-smoothable if for any $\mu > 0$, there exists a convex approximation f_μ such that:

1. $f_\mu(x) \leq f(x) \leq f_\mu(x) + \beta\mu$, for all x.
2. f_μ is $\frac{\alpha}{\mu}$-smooth.

We also call f_μ as the $\frac{1}{\mu}$-smooth approximation of f with parameters (α, β).

Example 15.3 (ℓ_1-Norm) We can approximate the absolute value $f(z) = |z|$ via the Huber loss

$$h_\mu(z) = \begin{cases} z^2/(2\mu), & \text{if } |z| \leq \mu; \\ |z| - \mu/2, & \text{otherwise.} \end{cases}$$

By Fig. 15.4, we can see that $h_\mu(z) \leq |z| \leq h_\mu(z) + \mu/2$. As $h_\mu(z)$ is $\frac{1}{\mu}$-smooth, so $|z|$ is $(1, \frac{1}{2})$-smoothable.

For the ℓ_1-norm $f(z) = \|z\|_1$, we can approximate it by $\sum_{i=1}^d h_\mu(z_i)$. Therefore, we have

$$\sum_{i=1}^d h_\mu(z_i) \leq \|z\|_1 \leq \sum_{i=1}^d h_\mu(z) + d\mu/2.$$

Therefore, $\|z\|_1$ is $(1, \frac{d}{2})$-smoothable.

Example 15.4 (ℓ_2-Norm) We can approximate the ℓ_2-norm $f(x) = \|x\|_2$ by

$$f_\mu(x) = \sqrt{\|x\|_2^2 + \mu^2} - \mu.$$

We have

$$f_\mu(x) \leq \|x\|_2 + \mu - \mu \leq \|x\|_2 \leq \sqrt{\|x\|_2^2 + \mu^2} = f_\mu(x) + \mu.$$

Thus, $\|x\|_2$ is $(1, 1)$-smoothable. Notice that compared to the ℓ_1-norm, the smoothing parameters of $\|x\|_2$ are dimension-free. Therefore, for the square-root loss $f(\beta) = \|Y - \mathbb{X}\beta\|_2$, we can approximate it by $f_\mu(\beta) = \sqrt{\|Y - \mathbb{X}\beta\|_2^2 + \mu^2} - \mu$. We can check that $f(\beta)$ is $(\|\mathbb{X}\|_{\text{op}}^2, 1)$-smoothable.

The following theorem shows the convergence rate using Nesterov's smoothing idea.

Theorem 15.1 *Given the objective function $F(x) = f(x) + h(x)$, where f is (α, β)-smoothable and h is convex, let f_μ be the $\frac{1}{\mu}$-smooth approximation to f. If we apply the accelerated proximal gradient descent to $F_\mu(x) = f_\mu(x) + h(x)$, where we choose $\mu = \epsilon/(2\beta)$, then*

$$F(x_t) - F(x^*) \leq \epsilon \text{ if } t \gtrsim \frac{\sqrt{\alpha\beta}}{\epsilon}.$$

Proof First, as f_μ is α/μ-smooth, Theorem 14.2 shows that we can achieve ϵ-accuracy

$$F_\mu(x_t) - \min_x F_\mu(x) \leq \epsilon$$

within $t = O(\sqrt{\frac{\alpha}{\mu}} \frac{1}{\sqrt{\epsilon}})$ steps. Second, by Definition 15.2, we have $|f(x_t) - f_\mu(x_t)| \leq \beta\mu = \frac{\epsilon}{2}$ as we choose $\mu = \epsilon/(2\beta)$. Therefore, we have

$$F(x_t) - F(x^*) \leq |f(x_t) - f_\mu(x_t)| + |F_\mu(x_t) - \min_x F_\mu(x)| \leq \epsilon$$

within $t = O(\sqrt{\frac{\alpha}{\mu}} \frac{1}{\sqrt{\epsilon}}) = O(\sqrt{\frac{\alpha\beta}{\epsilon}} \frac{1}{\sqrt{\epsilon}}) = O(\frac{\sqrt{\alpha\beta}}{\epsilon})$ steps. □

For the square-root Lasso, if we apply the accelerated proximal gradient descent to

$$\min_\beta \sqrt{\|Y - \mathbb{X}\beta\|_2^2 + \mu^2} + \lambda\|\beta\|_1.$$

As $f(\beta) = \|Y - \mathbb{X}\beta\|_2$ is $(\|\mathbb{X}\|_{\text{op}}^2, 1)$-smoothable, as long as the maximum singular value of the design matrix $\mathbb{X}$ is bounded, we can achieve ϵ-accuracy of square-root Lasso within $O(1/\epsilon)$ steps.

Another idea is why not we apply the Huber loss to approximation the ℓ_1-norm in Lasso. Instead of applying FISTA, we apply the accelerated gradient descent to

$$\min_\beta \|Y - \mathbb{X}\beta\|_2^2 + \lambda \sum_{j=1}^d h_\mu(\beta_j),$$

where $h_\mu(z)$ is defined in Example 15.3. As the ℓ_1-norm $\|\beta\|_1$ is $(1, \frac{d}{2})$-smoothable, we achieve ϵ-accuracy within $O(\frac{\sqrt{d}}{\epsilon})$ steps. When the dimension d is large, the convergence rate is not good.

Chapter 16
Duality and ADMM

16.1 Composite Objective Function

In the previous chapters, we introduce the proximal gradient descent to solve the optimization problem $\min_x f(x) + g(x)$, where f is smooth but g is not differentiable. A key step in the proximal gradient descent is the solve the proximal operator

$$\text{prox}_g(x) = \arg\min_{z \in \mathbb{R}^d} \left\{ \frac{1}{2} \|z - x\|_2^2 + g(z) \right\}.$$

For example, when $g(x) = \lambda \|x\|_1$, the proximal operator is soft-thresholding. For the fused Lasso problem

$$\min_\beta \|Y - \mathbb{X}\beta\|_2^2 + \lambda \|D\beta\|_1,$$

where D is the differential map (see Sect. 11.3.2), if we apply the proximal gradient descent to solve the above problem, we need to solve the sub-problem

$$\arg\min_{z \in \mathbb{R}^d} \left\{ \frac{1}{2} \|z - x\|_2^2 + \lambda \|Dz\|_1 \right\}.$$

Unlike the ℓ_1-norm, the above problem does not have a closed form solution. Therefore, it will be inefficient to implement proximal gradient descent to solve fused Lasso.

In this chapter, we will discuss how to solve this type of composite objective functions. In general, we are interested in the following composite optimization problem

$$\min_{x,y} f(x) + g(y), \text{ s.t. } Ax + By = c, \tag{16.1}$$

J. Lu, *Big Data Analysis*, https://doi.org/10.1007/978-3-032-03161-7_16

for some convex f and g. We can see that the fused Lasso can reduce the formality above by letting $f(\beta) = \|Y - \mathbb{X}\beta\|_2^2$, $g(y) = \lambda\|y\|_1$ so

$$\min_{\beta} \|Y - \mathbb{X}\beta\|_2^2 + \lambda\|D\beta\|_1 \Longleftrightarrow \min_{\beta,y} f(\beta) + g(y), \text{ s.t. } D\beta - y = 0.$$

Another example is the basis pursuit

$$\min_{\beta} \|\beta\|_1 \text{ s.t. } \mathbb{X}\beta - Y = 0.$$

16.2 Duality

In order to solve the composite optimization problem in (16.1), we start with reviewing the concept of duality in optimization. We define the Lagrange multiplier function of (16.1) as

$$L(x, y, \lambda) = f(x) + g(y) + \lambda^\top (Ax + By - c).$$

This allows us to convert the **primal problem** in (16.1) to the

$$\textbf{Dual problem} : \max_{\lambda} h(\lambda), \text{ where } h(\lambda) = \min_{x,y} L(x, y, \lambda).$$

Duality is an important concept in optimization. The following theorem implies that we can solve the primal problem via the dual problem.

Theorem 16.1 (Strong Duality) *Given convex functions f, g, the primal problem has the solution*

$$(x^*, y^*) = \mathop{arg\,min}_{x,y} f(x) + g(y), \text{ s.t. } Ax + By = c.$$

Consider the dual problem $\lambda^* = \text{argmax}_\lambda h(\lambda)$, where $h(\lambda) = \min_{x,y} L(x, y, \lambda)$. We can solve (x^*, y^*) via

$$(x^*, y^*) = \mathop{arg\,min}_{x,y} L(x, y, \lambda^*).$$

As $L(x, y, \lambda^*)$ is decomposable, we further have

$$x^* = \mathop{arg\,min}_{x} f(x) + \lambda^{*\top} Ax \text{ and } y^* = \mathop{arg\,min}_{y} g(y) + \lambda^{*\top} By.$$

The above strong duality property is useful when we want to convert a linear constraint problem to an unconstrained dual problem. The following example shows how we can find the dual problem of Lasso.

Example 16.1 (Duality of Lasso) As Lasso is an unconstrained problem, we first convert it to the standard formality in (16.1) as

$$\min_{\beta} \frac{1}{2}\|Y - \mathbb{X}\beta\|_2^2 + \lambda\|\beta\|_1 \iff \min_{\beta,z} \frac{1}{2}\|z\|_2^2 + \lambda\|\beta\|_1, \text{ s.t. } Y - \mathbb{X}\beta = z.$$

The Lagrange function is

$$L(\beta, z, \mu) = \frac{1}{2}\|z\|_2^2 + \lambda\|\beta\|_1 + \mu^\top(Y - \mathbb{X}\beta - z).$$

Notice that $L(\beta, z, \mu)$ is decomposable

$$\min_{\beta,z} L(\beta, z, \mu) = \min_{\beta} \left\{\lambda\|\beta\|_1 - \mu^\top\mathbb{X}\beta\right\} + \min_{z} \left\{\frac{1}{2}\|z\|_2^2 - \mu^\top z\right\} + \mu^\top Y.$$

The second problem above is quadratic, and we have

$$z^* = \arg\min_{z} \left\{\frac{1}{2}\|z\|_2^2 - \mu^\top z\right\} \implies z^* - \mu = 0. \tag{16.2}$$

We also apply the first-order optimality condition to the first problem and get

$$\beta^* = \arg\min_{\beta} \left\{\lambda\|\beta\|_1 - \mu^\top\mathbb{X}\beta\right\} \implies 0 = -\mathbb{X}^\top\mu + \lambda g, \text{ for some } g \in \partial\|\beta^*\|_1.$$

As $\|g\|_\infty \le 1$, the equality $0 = -\mathbb{X}^\top\mu + \lambda g$ has a finite solution only if $\lambda \ge \|\mathbb{X}^\top\mu\|_\infty$. Moreover, if $\lambda \ge \|\mathbb{X}^\top\mu\|_\infty$, we multiply β^* on the both sides of the equation $0 = -\mathbb{X}^\top\mu + \lambda g$ and have

$$0 = -\beta^{*\top}\mathbb{X}^\top\mu + \lambda\beta^{*\top}g = -(\beta^*)^\top\mathbb{X}^\top\mu + \lambda\|\beta^*\|_1,$$

where we use the fact that $\beta^{*\top}g = \beta^{*\top}\text{sign}(\beta^*) = \|\beta^*\|_1$ if $g \in \partial\|\beta^*\|_1$. Therefore, we have

$$\min_{\beta} \left\{\lambda\|\beta\|_1 - \mu^\top\mathbb{X}\beta\right\} = \begin{cases} 0 & \text{if } \lambda \ge \|\mathbb{X}^\top\mu\|_\infty; \\ -\infty & \text{if } \lambda < \|\mathbb{X}^\top\mu\|_\infty. \end{cases} \tag{16.3}$$

Combining(16.2) and (16.3), the dual problems become

$$\begin{aligned} \max_{\lambda} \min_{\beta,z} L(\beta, z, \mu) &= \max_{\lambda} \left\{ \min_{\beta} \left\{\lambda\|\beta\|_1 - \mu^\top\mathbb{X}\beta\right\} \right. \\ &\qquad \left. + \min_{z} \left\{\frac{1}{2}\|z\|_2^2 - \mu^\top z\right\} + \mu^\top Y \right\} \\ &= \max_{\lambda} -\frac{1}{2}\|\mu\|_2^2 + \mu^\top Y \text{ s.t. } \lambda \ge \|\mathbb{X}^\top\mu\|_\infty, \end{aligned}$$

as when $\lambda < \|\mathbb{X}^\top \mu\|_\infty$, the objective value of the dual becomes $-\infty$ which could not be the maximum. Therefore, the dual problem of Lasso is

$$\max_\mu L(\beta^*, z^*, \mu) = \max_\mu -\frac{1}{2}\|\mu\|_2^2 + \mu^\top Y \text{ s.t. } \|\mathbb{X}^\top \mu\|_\infty \leq \lambda.$$

The above problem is equivalent to

$$\textbf{Duality of Lasso}: \min_\mu \|Y - \mu\|_2^2, \text{ s.t. } \|\mathbb{X}^\top \mu\|_\infty \leq \lambda. \tag{16.4}$$

By (16.2), we can see that the dual variable $\mu^* = z^* = Y - \mathbb{X}\beta^*$ is the residual. So the duality of Lasso gives us a new geometric insight to Lasso. This duality form illustrates that, just like OLS, the Lasso is also a projection. Instead of projecting Y on the hyperplane $C(\mathbb{X})$ for OLS, the Lasso projects Y on the polytope $C(\lambda) = \{\mu | \ \|\mathbb{X}^\top \mu\|_\infty \leq \lambda\}$. The residual $\mu^* = Y - \mathbb{X}\beta^* = P_{C(\lambda)}(Y)$, where P_M is the projection operator such that $P_M(x) = \arg\min_{z\in M} \|z - x\|_2$.

If Y locates in the white region, i.e., $\|\mathbb{X}^\top Y\|_\infty \leq \lambda$, then $P_{C(\lambda)}(Y) = Y = Y - \mathbb{X}\beta^*$. This implies that $\beta^* = 0$ as all the other vector in null($\mathbb{X}$) has a larger ℓ_1-norm. If Y lies in the red region, the projection is to one of the faces of the polytope. If Y lies in the blue region, the projection is to one of the vertices of the polytope. If we shrink the value of λ, the projection of Y will remain on the same face until Y locates in the blue region and may further pivot to another red region, which leads to a piecewise linear regularization path of Lasso.

16.3 Alternating Direction Method of Multipliers

We consider an equivalent form of (16.1) by adding a quadratic term as follows:

$$\begin{cases} \min\limits_{x,y} f(x) + g(y) \\ \text{s.t. } Ax + By = c, \end{cases} \iff \begin{cases} \min\limits_{x,y} f(x) + g(y) + \frac{\rho}{2}\|Ax + By - c\|_2^2 \\ \text{s.t. } Ax + By = c. \end{cases}$$

We then introduce **augmented Lagrangian** for the second problem above:

$$L_\rho(x, y, \lambda) = f(x) + g(y) + \lambda^\top (Ax + By - c) + \frac{\rho}{2}\|Ax + By - c\|_2^2.$$

In order to solve the dual problem $\max_\lambda \min_{x,y} L_\rho(x, y, \lambda)$, we propose to update the primal variables x, y and dual variables alternatively as follows:

$$\text{Primal step: } \begin{cases} x_{t+1} = \arg\min_x L_\rho(x, y_t, \lambda_t); \\ y_{t+1} = \arg\min_y L_\rho(x_{t+1}, y, \lambda_t); \end{cases}$$

$$\textbf{Dualstep}: \quad \lambda_{t+1} = \arg\max_{\lambda} L_\rho(x_{t+1}, y_{t+1}, \lambda) - \underbrace{\frac{1}{2\rho}\|\lambda - \lambda_t\|_2^2}_{\text{Proximal term}}.$$

We add a proximal term in the dual step above, because the augmented Lagrangian $L_\rho(x_{t+1}, y_{t+1}, \lambda)$ is linear with respect to λ and thus $\max_\lambda L_\rho(x_{t+1}, y_{t+1}, \lambda) = \infty$. So we need to add a quadratic proximal term to stop λ_{t+1} deviating from λ_t too much.

Plugging the augmented Lagrangian into the above primal and dual steps, we have the following algorithm.

The **alternating direction method of multipliers (ADMM)** solves (16.1) via

$$x_{t+1} = \arg\min_x f(x) + \frac{\rho}{2}\|Ax + By_t - c + \lambda_t/\rho\|_2^2$$

$$y_{t+1} = \arg\min_y g(x) + \frac{\rho}{2}\|Ax_{t+1} + By - c + \lambda_t/\rho\|_2^2$$

$$\lambda_{t+1} = \lambda_t + \rho(Ax_{t+1} + By_{t+1} - c).$$

It can be shown that if f and g are closed convex functions,[1] ADMM algorithm converges.

Example 16.2 (Fused Lasso) We apply ADMM to the fused Lasso

$$\min_\beta \frac{1}{2}\|Y - \mathbb{X}\beta\|_2^2 + \lambda\|D\beta\|_1 \iff \min_{\beta,z} \frac{1}{2}\|Y - \mathbb{X}\beta\|_2^2 + \lambda\|z\|_1 \text{ s.t. } D\beta - z = 0,$$

and get the updating rule as

$$\beta_{t+1} = \arg\min_\beta \frac{1}{2}\|Y - \mathbb{X}\beta\|_2^2 + \frac{\rho}{2}\|D\beta - z_t + \lambda_t/\rho\|_2^2$$

$$z_{t+1} = \arg\min_z \lambda\|z\|_1 + \frac{\rho}{2}\|D\beta_{t+1} - z + \lambda_t/\rho\|_2^2$$

$$\lambda_{t+1} = \lambda_t + \rho(D\beta_{t+1} - z_{t+1}).$$

This further yields the algorithm

[1] A convex function f is closed if $\{x \mid f(x) \leq \alpha\}$ is a closed set for any α.

$$\beta_{t+1} = (\mathbb{X}^\top \mathbb{X} + \rho D^\top D)^{-1}(\mathbb{X}^\top Y + \rho D^\top z_t - D^\top \lambda_t)$$
$$z_{t+1} = \text{SoftThreshold}(D\beta_{t+1} + \lambda_t/\rho, \lambda/\rho)$$
$$\lambda_{t+1} = \lambda_t + \rho(D\beta_{t+1} - z_{t+1}).$$

When $D = I$, the above algorithm reduces to the ADMM algorithm for Lasso. From these examples, we can also see that ADMM is not a first-order method as it involves information of f and g. For Lasso, unlike FISTA, the ADMM algorithm involves the matrix inversion which may be time-consuming when d is large.

Example 16.3 (Graphical Lasso) Given i.i.d. samples $X_1, \ldots, X_n \sim N(0, \Sigma)$, we aim to estimate the precision matrix $\Theta = \Sigma^{-1}$ via the graphical Lasso (see Chap. 11):

$$\min_{\Theta} - \log\det\Theta + \text{tr}(\Theta^\top \widehat{\Sigma}) + \lambda\|\Theta\|_{1,1}$$

$$\Updownarrow$$

$$\min_{\Theta,\Psi} - \log\det\Theta + \text{tr}(\Theta^\top \widehat{\Sigma}) + \lambda\|\Psi\|_{1,1} \text{ s.t. } \Theta = \Psi.$$

Applying the ADMM algorithm we get

$$\Theta_{t+1} = \arg\min_{\Theta} - \log\det\Theta + \text{tr}(\Theta^\top \widehat{\Sigma}) + \frac{\rho}{2}\|\Theta - \Psi_t + \Lambda_t/\rho\|_F^2$$
$$\Psi_{t+1} = \arg\min_{\Psi} \lambda\|\Psi\|_{1,1} + \frac{\rho}{2}\|\Theta_{t+1} - \Psi + \Lambda_t/\rho\|_F^2$$
$$\Lambda_{t+1} = \Lambda_t + \rho(\Theta_{t+1} - \Psi_{t+1}),$$

where the Frobenious norm $\|A\|_F^2 = \sum_{jk} A_{jk}^2$ and it can be further simplified as

$$\Theta_{t+1} = \mathcal{F}_\rho(\Psi_t - \Lambda_t/\rho - \widehat{\Sigma}/\rho)$$
$$\Psi_{t+1} = \text{SoftThreshold}(\Theta_{t+1} + \Lambda_t/\rho, \lambda/\rho)$$
$$\Lambda_{t+1} = \Lambda_t + \rho(\Theta_{t+1} - \Psi_{t+1}),$$

where for the spectral decomposition $X = UDU^\top$, where $D = \text{diag}(\lambda_1, \ldots, \lambda_n)$, one has $\mathcal{F}_\rho(X) = U\text{diag}\{\lambda_i + \sqrt{\lambda_i + 4/\rho}\}U^\top$.

The last example shows that ADMM algorithm can be implemented as a distributed algorithm.

Example 16.4 (Consensus Optimization) We aim to design a divide-and-conquer ADMM for Lasso for massive data when n is ultra-large. The Lasso

$$\min_{\beta} \sum_{i=1}^{n} (Y_i - \mathbb{X}_i \beta)^2 + \lambda \|\beta\|_1$$

can be formulated as the general block separable form

$$\min_{x} \sum_{i=1}^{n} f_i(x) \iff \min_{x_i, z} \sum_{i=1}^{n} f_i(x) \text{ s.t. } x_i = z \text{ for all } i = 1, \ldots, n.$$

Applying the ADMM algorithm we get

$$\begin{aligned}
\text{Divide} : x_i^{t+1} &= \arg\min_{x_i} f_i(x_i) + \frac{\rho}{2} \|x_i - z^t + \lambda_i^t/\rho\|_2^2, \forall i = 1, \ldots, n \\
\text{Gather} : z^{t+1} &= \frac{1}{n} \sum_{i=1}^{n} (x_i^{t+1} + \lambda_i^t/\rho) \\
\text{Broadcast} : \lambda_i^{t+1} &= \lambda_i^t + \rho(x_i^{t+1} - z^{t+1}), \forall i = 1, \ldots, n.
\end{aligned}$$

In the first step, we can update each x_i^{t+1} in parallel, then in the second step, we gather all local iterates, and in the third step, we broadcast the updated dual variables back to each core.

Part V
High-Dimensional Inference

Chapter 17
High Dimensional Inference

17.1 Statistical Inference

We start the part of high-dimensional inference by introducing the problems in statistical inference. Given the statistical model $\{\mathbb{P}_\theta | \theta \in \Theta\}$, we observe $X_1, \ldots, X_n \overset{iid}{\sim} \mathbb{P}_{\theta^*}$ where θ^* is the truth. Here are the major goals to estimation and inference.

- **Estimation**: We want to find an estimator $\widehat{\theta}$ and we're interested in the estimation rate or the statistical rate $\|\widehat{\theta} - \theta^*\| = O_P(r_n)$ for some rate r_n. In terms of estimation, we'd like to know whether the estimator converges to the truth in probability or how fast it converges to the truth in probability. For example, we show that under certain conditions, Lasso estimator has the ℓ_2-norm rate $O_P(\sqrt{s \log d/n})$.
- **Inference**: We aim to pursue more than a consistent estimator. The inference problems focus on the uncertainty assessment of our estimates. We aim to measure how much certainty we have to our estimates, how significant are the detected signals, and how confident are our statistical decisions.

In statistical inference, there are two major problems: confidence interval and hypothesis testing.

- **Confidence interval**: We aim to construct a confidence interval I_α such that the true parameter belongs to the confidence interval with probability larger than $1 - \alpha$:

$$\mathbb{P}(\theta^* \in I_\alpha) \geq 1 - \alpha.$$

- **Hypothesis testing**: We aim to test certain properties of the true parameters, whether the true parameters belong to the null space Θ_0 or it belongs to the alternative Θ_1:

J. Lu, *Big Data Analysis*, https://doi.org/10.1007/978-3-032-03161-7_17

$$H_0 : \theta^* \in \Theta_0 \quad v.s. \quad H_1 : \theta^* \in \Theta_1.$$

We aim to propose a test with the significance level $1 - \alpha$

$$\psi_\alpha = \begin{cases} 0, & \text{not reject } H_0; \\ 1, & \text{reject } H_0, \end{cases}$$

such that the type-I error $\mathbb{P}_{H_0}(\psi_\alpha = 1) \leq \alpha$ and the type-II error $\mathbb{P}_{H_1}(\psi_\alpha = 0) \to 1$ as $n \to \infty$.

17.2 High-Dimensional Inference

In high-dimensional inference, our parameters of interest $\theta^* \in \mathbb{R}^d$ are typically large, for most cases, larger than the sample sizes n. What will be unique for high-dimensional inference?

The first thing is that instead of testing single hypothesis, we will test multiple hypotheses. For example, let each entry of θ^* represent the effect of each gene and we want to select which genes are effective from our data. Instead of testing a single gene, we need to test millions or even trillions of genes. We need to consider the following multiple hypotheses:

$$H_{0j} : \theta_j^* \in \Theta_0 \quad v.s. \quad H_{1j} : \theta_j^* \in \Theta_1 \qquad \forall j = 1, \ldots, N.$$

Conducting the multiple hypotheses is much harder than a single hypothesis due to the multiplicity. Here is a story from John Tukey.

A young psychologist administers 250 hypothesis tests as part of a research project and finds that 11 were significant at the 5% level. The young researcher feels very proud of this fact and is ready to make a big deal about it, until a senior researcher (Tukey himself?) suggests that one would expect 12.5 significant tests even in the purely null case, merely by chance, because

$$250 \times 5\% = 12.5.$$

In that sense, finding only 11 significant results is actually somewhat disappointing!

Tukey's story tells us the lesson that we should be cautious when conducting multiple hypothesis. The following criterion would be one way to control the multiplicity in the hypotheses.

Definition 17.1 (Family-Wise Error Rate) The family-wise error rate (FWER) is the probability of making at least one type-I error in the multiple hypotheses

$$\text{FWER} = \mathbb{P}(\text{ Reject any true } H_{0j}) = \mathbb{P}(\psi_j = 1 \text{ for some } j \in H_0).$$

Family-wise error rate might be too stringent that we are not allowed to make any type-I errors. Another widely used criterion is called the false discovery rate.

Definition 17.2 (False Discovery Rate) Define the false discovery proportion (FDP) as the proportion of false rejections among all rejections:

$$\mathrm{FDP} = \sum_{j \in H_0} \frac{\psi_j}{\max\{\sum_{j=1}^{d} \psi_j, 1\}},$$

where we add 1 to avoid the denominator being 0. The false discovery rate is the expected FDP, i.e., FDR $= \mathbb{E}[\mathrm{FDP}]$.

We will talk about how to conduct multiple hypotheses in the following chapters.

17.3 Asymptotic Normality of Least Squares

We first review the important theoretical results for inference. The cornerstone of inference is the central limit theorem.

Theorem 17.1 (Central Limit Theorem) *Let $X_1, \ldots, X_n, \ldots$ be i.i.d. random variables with $\mathbb{E}(X_i) = \mu$ and $\mathrm{Var}(X_i) = \sigma^2 < \infty$ for $i = 1, 2, \ldots$. The sample mean $\bar{X}_n = (X_1 + \ldots + X_n)/n$ converges in distribution to the normal distribution. In particular, we have*

$$\sqrt{n}\left(\frac{\bar{X}_n - \mu}{\sigma}\right) \xrightarrow{d} N(0, 1).$$

A typical inference pipeline for hypothesis: first, we construct the statistics T for H_0 and then we want to show that T can be decomposed as

$$T = \underbrace{\frac{1}{\sqrt{n}} \sum_{i=1}^{n} (g(X_i) - \mathbb{E}[g(X)])}_{\text{Leading term}} + \underbrace{o_P(1)}_{\text{Remainder}}$$

where the leading term yields the asymptotic normality by CLT and the remainder term is small enough so it is ignorable. The following theorem is useful when we have such decomposition.

Theorem 17.2 (Slutsky's Theorem) *If a random variable $X_n \xrightarrow{d} X$ and another random variable $Y_n \xrightarrow{P} c$ where c is a constant, then*

$$X_n + Y_n \xrightarrow{d} X + c,\ X_n Y_n \xrightarrow{d} cX,\ \text{and } X_n/Y_n \xrightarrow{d} X/c.$$

Now we are ready to derive the asymptotic normality of the ordinary least squares.

Theorem 17.3 (Asymptotic Normality for OLS) *The linear model* $Y = \mathbb{X}\beta^* + \epsilon$ *has the design matrix* $\mathbb{X} \in \mathbb{R}^{n\times d}$ *with* $\mathrm{rank}(\mathbb{X}) = d \le n$ *and the i.i.d. centered noise* $Var(\varepsilon_i) = \sigma^2$. *The ordinary least squares estimator is*

$$\widehat{\beta}^{\mathrm{LS}} = (\mathbb{X}^\top\mathbb{X})^{-1}\mathbb{X}^\top Y.$$

Denote sample covariance matrix $\widehat{\Sigma} = \frac{1}{n}\mathbb{X}^\top\mathbb{X}$, *then we have*

$$\begin{aligned}\sqrt{n}(\widehat{\beta}^{\mathrm{LS}} - \beta^*) &= \sqrt{n}[(\mathbb{X}^\top\mathbb{X})^{-1}\mathbb{X}^\top(\mathbb{X}\beta^* + \varepsilon) - \beta^*] \\ &= \sqrt{n}(\mathbb{X}^\top\mathbb{X})^{-1}\mathbb{X}^\top\varepsilon \\ &= \frac{1}{\sqrt{n}}\sum_{i=1}^{n}\widehat{\Sigma}^{-1}X_i\varepsilon_i \xrightarrow{d} N(0, \sigma^2\widehat{\Sigma}^{-1}).\end{aligned}$$

Notice that we assume the fixed design so the asymptotic normality results above are conditioned on $\mathbb{X}$.

Chapter 18
Debiased Lasso

18.1 Debiased Lasso

In this chapter, we aim to conduct inference for high-dimensional linear model. Recall the sparse linear model $Y = \mathbb{X}\beta^* + \varepsilon$, where $\mathbb{X} \in \mathbb{R}^{n\times d}$ and $\|\beta^*\|_0 \le s$. We aim to derive the confidence interval for the Lasso estimator

$$\widehat{\beta} = \arg\min_{\beta} \frac{1}{2n}\|Y - \mathbb{X}\beta\|_2^2 + \lambda\|\beta\|_1.$$

By the first-order optimality condition, we have

$$-\frac{1}{n}\mathbb{X}^\top(Y - \mathbb{X}\widehat{\beta}) + \lambda z = 0, \text{ where } z \in \partial\|\widehat{\beta}\|_1. \tag{18.1}$$

Plug the model $Y = \mathbb{X}\beta^* + \varepsilon$ into (18.1) and denote the sample covariance matrix $\widehat{\Sigma} = \mathbb{X}^\top\mathbb{X}/n$; we have

$$\widehat{\Sigma}(\widehat{\beta} - \beta^*) + \lambda z = \frac{1}{n}\mathbb{X}^\top\varepsilon. \tag{18.2}$$

We expect to "inverse" the sample covariance matrix above; however, when $d > n$, $\widehat{\Sigma}$ is singular. Our idea is that suppose we have a matrix $\widehat{\Theta}$ which is close enough to the inverse, i.e., $\widehat{\Theta}\widehat{\Sigma} \approx I$, we can reveal the parameter in (18.2). Multiplying $\widehat{\Theta}$ on both sides of (18.2) and rearranging the equation, we have

$$\sqrt{n}(\widehat{\beta} - \beta^* + \underbrace{\lambda\widehat{\Theta}z}_{\text{Bias}}) = \underbrace{\frac{1}{\sqrt{n}}\widehat{\Theta}\mathbb{X}^\top\varepsilon}_{\text{Leading term}} + \underbrace{\sqrt{n}(I - \widehat{\Theta}\widehat{\Sigma})(\widehat{\beta} - \beta^*)}_{\text{Remainder}}. \tag{18.3}$$

J. Lu, *Big Data Analysis*, https://doi.org/10.1007/978-3-032-03161-7_18

We can see that the subgradient in the above equation is a bias, and we successfully decompose the right-hand side into the leading term and remainder term. By (18.1), we have $\lambda z = \frac{1}{n}\mathbb{X}^\top(Y - \mathbb{X}\widehat{\beta})$ and plug it into (18.3), then

$$\sqrt{n}\left(\widehat{\beta} + \frac{1}{n}\widehat{\Theta}\mathbb{X}^\top(Y - \mathbb{X}\widehat{\beta}) - \beta^*\right) = \frac{1}{\sqrt{n}}\widehat{\Theta}\mathbb{X}^\top\varepsilon + \sqrt{n}(I - \widehat{\Theta}\widehat{\Sigma})(\widehat{\beta} - \beta^*). \tag{18.4}$$

We can define the **debiased Lasso**

$$\widehat{\beta}^d = \widehat{\beta} + \frac{1}{n}\widehat{\Theta}\mathbb{X}^\top(Y - \mathbb{X}\widehat{\beta})$$

and then we have

$$\sqrt{n}(\widehat{\beta}^d - \beta^*) = \frac{1}{\sqrt{n}}\widehat{\Theta}\mathbb{X}^\top\varepsilon + \sqrt{n}(I - \widehat{\Theta}\widehat{\Sigma})(\widehat{\beta} - \beta^*).$$

We have following asymptotic normality of the debiased Lasso.

Theorem 18.1 (Asymptotic Normality of Debiased Lasso) *The linear model $Y = \mathbb{X}\beta^* + \varepsilon$ has the i.i.d. random design $\mathbb{X} = (X_1, \ldots, X_n)^\top$ with positive definite covariance $Cov(X_i) = \Sigma$ and the i.i.d. centered noise has $Var(\varepsilon_i) = \sigma^2$. We assume both X_{ij} for all $j = 1, \ldots, d$ and ε_i are sub-Gaussian. Suppose we can find an estimator $\widehat{\Theta}$ of the precision matrix $\Theta = \Sigma^{-1}$ satisfying the following conditions:*

$$\textbf{C1.} \|\widehat{\beta} - \beta^*\|_1 = O_P(s\sqrt{\log d/n});$$

$$\textbf{C2.} \|\widehat{\Sigma}\widehat{\Theta} - I\|_{\max} = O_P(\sqrt{\log d/n});$$

$$\textbf{C3.} \max_{1\le j\le d} \|\widehat{\Theta}_j - \Theta_j\|_1 = O_P(s\sqrt{\log d/n}).$$

If $s \log d/\sqrt{n} = o(1)$, we have

$$\sqrt{n}(\widehat{\beta}^d_j - \beta^*_j) \xrightarrow{d} N(0, \Theta_{jj}), \text{ for all } j = 1, \ldots, d.$$

Proof By (18.4), we have

$$\sqrt{n}(\widehat{\beta}^d - \beta^*) = \text{Leading Term} + \text{Remainder},$$

where Leading Term $= \frac{1}{\sqrt{n}}\widehat{\Theta}\mathbb{X}^\top\varepsilon$ and Remainder $= \sqrt{n}(I - \widehat{\Theta}\widehat{\Sigma})(\widehat{\beta} - \beta^*)$. We first bound the remainder term by

$$\begin{aligned}\|\text{Remainder}\|_\infty &= \|\sqrt{n}(I - \widehat{\Theta}\widehat{\Sigma})(\widehat{\beta} - \beta^*)\|_\infty \\ &\leq \sqrt{n}\|\widehat{\Sigma}\widehat{\Theta} - I\|_{\max}\|\widehat{\beta} - \beta^*\|_1 = O_P(s\log d/\sqrt{n}) = o_P(1),\end{aligned}$$

where in the first inequality, we use the matrix Hölder inequality $\|Ax\|_\infty \leq \|A\|_{\max}\|x\|_1$ and, in the first O_P, we use the conditions C1 and C2. We decompose the leading term as

$$\text{Leading Term} = \frac{1}{\sqrt{n}}\widehat{\Theta}\mathbb{X}^\top\varepsilon = \underbrace{\frac{1}{\sqrt{n}}\Theta\mathbb{X}^\top\varepsilon}_{L_1} + \underbrace{\frac{1}{\sqrt{n}}(\widehat{\Theta} - \Theta)\mathbb{X}^\top\varepsilon}_{L_2}.$$

We have

$$\begin{aligned}\|L_2\|_\infty &= \|\frac{1}{\sqrt{n}}(\widehat{\Theta} - \Theta)\mathbb{X}^\top\varepsilon\|_\infty \\ &\leq \sqrt{n}\max_{1\leq j\leq d}\|\widehat{\Theta}_j - \Theta_j\|_1\|\frac{1}{n}\mathbb{X}^\top\varepsilon\|_\infty = O_P(s\log d/\sqrt{n}) = o_P(1),\end{aligned}$$

where we use Hölder inequality in the first inequality and apply C3 and Corollary 10.1. By CLT, the first term

$$L_1 = \frac{1}{\sqrt{n}}\Theta\mathbb{X}^\top\varepsilon \xrightarrow{d} N(0, \Theta).$$

As $\sqrt{n}(\widehat{\beta}^d - \beta^*) = L_1 + L_2 + \text{Remainder}$ and $\|L_2 + \text{Remainder}\|_\infty = o_P(1)$, by Slutsky's theorem, we have

$$\sqrt{n}(\widehat{\beta}_j^d - \beta_j^*) \xrightarrow{d} N(0, \Theta_{jj})$$

for all $j = 1, \ldots, d$. □

There are a few important observations from Theorem 18.1.

- **Regularization**: We did not use the fact that the regularization is the ℓ_1-norm in the proof. Notice that if we change the regularization to other penalties, (18.1) still holds along with the remaining derivations. Theorem 18.1 is true for general penalties.
- **Estimator**: If we change the penalty, the only condition on the new estimator $\widehat{\beta}$ we need to verify is C1 in Theorem 18.1. We will check it for Lasso soon. If the estimator using other penalties satisfies C1, we will still have the asymptotic normality.

- $\widehat{\Theta}$: Notice $\widehat{\Theta}$ is irrelevant to the estimation. We need it only because we want to derive asymptotic normality. So as long as you can find some estimator $\widehat{\Theta}$ satisfying C2 and C3, Theorem 18.1 holds.

We first check C1 in Theorem 18.1 which is related to the rate of Lasso.

Theorem 18.2 **(ℓ_1-Norm Rate of Lasso)** *Under the same conditions as the Corollary 10.1, we have*

$$\|\widehat{\beta} - \beta^*\|_1 = O_P(s\sqrt{\log d/n}).$$

Proof Corollary 10.1 shows the ℓ_2-norm of Lasso $\|\widehat{\beta} - \beta^*\|_2 = O_P(\sqrt{s \log d/n})$. In the proof of Corollary 10.1, we show that $\Delta = \widehat{\beta} - \beta^* \in \mathbb{C}_3(S) = \{\Delta | \|\Delta_{S^c}\|_1 \leq 3\|\Delta_S\|_1\}$. Therefore, we have

$$\|\Delta\|_1 = \|\Delta_S\|_1 + \|\Delta_{S^c}\|_1 \leq 4\|\Delta_S\|_1 \leq 4\sqrt{s}\|\Delta\|_2 = O_P(s\sqrt{\log d/n}).$$

□

18.2 CLIME Estimator

In this section, we will provide a concrete estimator $\widehat{\boldsymbol{\Theta}}$ satisfying C2 and C3 in Theorem 18.1.

The **constrained ℓ_1-minimization for inverse matrix estimation (CLIME)** estimates the jth column of the precision matrix Θ_j via

$$\widehat{\Theta}_j = \arg\min_{\beta} \|\beta\|_1 \text{ s.t. } \|\widehat{\Sigma}\beta - e_j\|_\infty \leq \lambda \tag{18.5}$$

for $j = 1, \ldots, d$, where λ is the tuning parameter and $e_j = (0, \ldots, 0, 1, 0, \ldots, 0)^\top$ is the jth canonical basis.

In history, the CLIME estimator was proposed to estimate a sparse precision matrix, i.e., the Gaussian graphical model (see Chap. 11). The ℓ_1-norm objective function finds a sparse column of $\widehat{\Theta}$ and the constraint is based on $\widehat{\Sigma}\widehat{\Theta}_j \approx e_j$ as we expect $\widehat{\Theta} \approx \Theta = \Sigma^{-1}$ and $\widehat{\Theta}\widehat{\Sigma} \approx I$.

In order to make sure that (18.5) is a reasonable estimator, we first need to check if the constraint is feasible.

Lemma 18.1 (Feasibility of CLIME) *If* $\max_j \|\Theta_j\|_1 < \infty$ *and we choose* $\lambda = C\sqrt{\log d/n}$ *for some sufficiently large constant* C*, then* $\max_j \|\widehat{\Sigma}\Theta_j - e_j\|_\infty \le \lambda$ *with probability at least* $1 - 1/d$*, i.e.,* Θ_j *is a feasible solution of the constraint of CLIME for all* $1 \le j \le d$.

Proof By definitions, we have $\Sigma\Theta = I$ and thus

$$\begin{aligned}\max_j \|\widehat{\Sigma}\Theta_j - e_j\|_\infty &= \max_j \|(\widehat{\Sigma} - \Sigma)\Theta_j\|_\infty \\ &\le \|\widehat{\Sigma} - \Sigma\|_{\max} \max_j \|\Theta_j\|_1 = O_P(\sqrt{\log d/n}), \end{aligned} \tag{18.6}$$

where in the last equality we use the condition $\max_j \|\Theta_j\|_1 < \infty$ and the result $\|\widehat{\Sigma} - \Sigma\|_{\max} = O_P(\sqrt{\log d/n})$ showed in the proof of Proposition 10.1. □

By the definition of CLIME, $\widehat{\Theta}_j$ must be a feasible solution to (18.5) for all $j = 1, \ldots, d$. Therefore, if we choose $\lambda = C\sqrt{\log d/n}$ for some sufficiently large constant C, we have C2 checked that

$$\|\widehat{\Sigma}\widehat{\Theta} - I\|_{\max} = \max_j \|\widehat{\Sigma}\widehat{\Theta}_j - e_j\|_\infty \le \lambda = O_P(\sqrt{\log d/n}).$$

What remained to be shown is C3 in Theorem 18.1, and it follows from the following theorem on the ℓ_1-norm rate of CLIME.

Theorem 18.3 (ℓ_1-Norm Rate of CLIME) *If* $\max_j \|\Theta_j\|_1 < \infty$ *and* $s = \max_j \|\Theta_j\|_0$ *and we choose* $\lambda = C\sqrt{\log d/n}$ *for some sufficiently large constant* C*, then*

$$\max_j \|\widehat{\Theta}_j - \Theta_j\|_1 = O_P(s\sqrt{\log d/n}).$$

Proof Define $\Delta_j = \widehat{\Theta}_j - \Theta_j$ and S is the support of Θ_j. As we show in Lemma 18.1 that Θ_j is a feasible solution to (18.5), so we have

$$0 \le \|\Theta_j\|_1 - \|\widehat{\Theta}_j\|_1 = \|\Theta_{jS}\|_1 - \|\widehat{\Theta}_{jS}\|_1 - \|\widehat{\Theta}_{jS^c}\|_1 \le \|\Delta_{jS}\|_1 - \|\Delta_{jS^c}\|_1.$$

Therefore, $\|\Delta_{jS^c}\|_1 \le \|\Delta_{jS}\|_1$. The proof is exactly the same as how we prove the Lasso error vector belongs to the cone. Now we first derive the ℓ_∞-norm of CLIME. By Lemma 18.1, we choose sufficiently large C for $\lambda = C\sqrt{\log d/n}$, so we have

$$\max_j \|\Theta_j\|_1 \|\widehat{\Sigma} - \Sigma\|_{\max} \le \lambda \tag{18.7}$$

Therefore, we have

$$\|\widehat{\Sigma}(\widehat{\Theta} - \Theta)\|_{\max} \le \|\widehat{\Sigma}\widehat{\Theta} - I\|_{\max} + \|\widehat{\Sigma}\Theta - I\|_{\max} \le 2\lambda, \tag{18.8}$$

where the second inequality is due to (18.6) and the definition of CLIME. We further have

$$\begin{aligned}\|\Sigma(\widehat{\Theta}-\Theta)\|_{\max} &\le \|\widehat{\Sigma}(\widehat{\Theta}-\Theta)\|_{\max} + \|(\widehat{\Sigma}-\Sigma)(\widehat{\Theta}-\Theta)\|_{\max} \\ &\le 2\lambda + \max_j \|\widehat{\Theta}_j - \Theta_j\|_\infty \|\widehat{\Sigma}-\Sigma\|_{\max} \\ &\le 2\lambda + 2\max_j \|\Theta_j\|_\infty \|\widehat{\Sigma}-\Sigma\|_{\max} \le 4\lambda,\end{aligned}$$

where the second inequality is by (18.8) and Hölder inequality, the third inequality is by the triangle inequality, $\|\Theta_j\|_1 \ge \|\widehat{\Theta}_j\|_1$, and the last inequality is by (18.7).

So we have the ℓ_∞-norm rate

$$\|\widehat{\Theta}-\Theta\|_{\max} \le \max_j \|\Theta_j\|_\infty \|\Sigma(\widehat{\Theta}-\Theta)\|_{\max} \le 4\max_j \|\Theta_j\|_\infty \lambda = O_P(\sqrt{\log d/n}).$$

As $\|\Delta_{jS^c}\|_1 \le \|\Delta_{jS}\|_1$ for all $j = 1, \ldots, d$, we have

$$\begin{aligned}\max_j \|\Delta_j\|_1 &= \max_j \|\Delta_{jS}\|_1 + \max_j \|\Delta_{jS^c}\|_1 \\ &\le \max_j 2\|\Delta_{jS}\|_1 \le 2s\max_j \|\Delta_{jS}\|_\infty = 2s\|\widehat{\Theta}-\Theta\|_{\max} \\ &= O_P(s\sqrt{\log d/n}),\end{aligned}$$

which completes the proof. □

18.2.1 Comparison of Lasso and Debiased Lasso

In order to have the asymptotic normality for high-dimensional linear model, we need to impose much stronger assumptions than the estimation. We need (1) the precision matrix of the design $\Theta = \Sigma^{-1}$ being sparse and (2) the scaling condition $\frac{s\log d}{\sqrt{n}} = o(1)$, neither of which is required to guarantee the consistency of Lasso.

18.3 General M-Estimator

In this section, we generalize the debiasing method to general high-dimensional M-estimators. For the loss $\mathcal{L}_n(\theta) = \frac{1}{n}\sum_{i=1}^n \ell(\theta, X_i)$, we assume that $\mathbb{E}[\nabla_\theta \ell(\theta^*, X)] = 0$, i.e., the truth θ^* is the minimizer of the population loss. We first consider the unregularized M-estimator

$$\widehat{\theta} = \arg\min_{\theta} \frac{1}{n} \sum_{i=1}^{n} \ell(\theta, X_i).$$

For example, for the logistic regression,

$$\ell(\theta, (Y, X)) = \log\big(1 + \exp(-YX^\top\theta)\big).$$

Applying the first-order optimality conditions, we would have

$$0 = \nabla_\theta \mathcal{L}_n(\widehat{\theta}) \approx \nabla_\theta \mathcal{L}_n(\theta^*) + \nabla_\theta^2 \mathcal{L}_n(\theta^*)(\widehat{\theta} - \theta^*)\widehat{\theta} - \theta^* \approx -[\nabla_\theta^2 \mathcal{L}_n(\theta^*)]^{-1} \nabla_\theta \mathcal{L}_n(\theta^*).$$

Therefore, we have

$$\sqrt{n}(\widehat{\theta} - \theta^*) \approx -\frac{1}{\sqrt{n}} \sum_{i=1}^{n} [\nabla_\theta^2 \mathcal{L}_n(\theta^*)]^{-1} \nabla_\theta \ell(\theta, X_i) \xrightarrow{d} N(0, ABA),$$

where $A = \mathbb{E}[\nabla_\theta^2 \ell(\theta^*, X)]^{-1}$, and $B = \mathbb{E}[\nabla_\theta \ell(\theta^*, X) \nabla_\theta \ell(\theta^*, X)^\top]$. By the property of Fisher information, we have $B^{-1} = -A$ and therefore

$$\sqrt{n}(\widehat{\theta} - \theta^*) \xrightarrow{d} N(0, \mathbb{E}[\nabla_\theta^2 \ell(\theta^*, X)]^{-1}).$$

When $\ell(\theta, X)$ is the log-likelihood, the asymptotic covariance $\mathbb{E}[\nabla_\theta^2 \ell(\theta^*, X)]^{-1}$ is the Fisher information.

Now we turn to the regularized M-estimator

$$\widehat{\theta} = \arg\min_{\theta} \mathcal{L}_n(\theta) + \lambda \|\theta\|_1. \tag{18.9}$$

By the first-order optimality conditions,

$$\nabla_\theta \mathcal{L}_n(\widehat{\theta}) + \lambda z = 0.$$

We expand $\nabla_\theta \mathcal{L}_n(\widehat{\theta})$ like above and get

$$\nabla_\theta \mathcal{L}_n(\theta^*) + \nabla_\theta^2 \mathcal{L}_n(\theta^*)(\widehat{\theta} - \theta^*) + \lambda z = 0.$$

Following the same strategy of Lasso, we aim to approximate the inversion of $\nabla_\theta^2 \mathcal{L}_n(\theta^*)$ via $\widehat{\Theta}$ and

$$\sqrt{n}(\widehat{\theta} - \widehat{\Theta} \nabla_\theta \mathcal{L}_n(\widehat{\theta}) - \theta^*) = -\frac{1}{\sqrt{n}} \sum_{i=1}^{n} \widehat{\Theta} \nabla_\theta \ell(\theta, X_i) + \sqrt{n}(I - \widehat{\Theta} \nabla_\theta^2 \mathcal{L}_n(\theta^*))(\widehat{\theta} - \theta^*).$$

The debiased M-estimator of (18.9) is

$$\widehat{\theta}^d = \widehat{\theta} - \widehat{\Theta}\nabla_\theta \mathcal{L}_n(\widehat{\theta}),$$

where we estimate the jth column of the inverse of Hessian matrix via

$$\widehat{\Theta}_j = \arg\min_{\beta} \|\beta\|_1 \text{ s.t. } \|\nabla_\theta^2 \mathcal{L}_n(\widehat{\theta})\beta - e_j\|_\infty \leq \lambda,$$

for $j = 1, \ldots, d$. Under the assumption that the Fisher information $\mathbb{E}[\nabla_\theta^2 \ell(\theta^*, X)]^{-1}$ is sparse and some other regularity conditions, we have

$$\sqrt{n}(\widehat{\theta}_j^d - \theta_j^*) \xrightarrow{d} N(0, (\mathbb{E}[\nabla_\theta^2 \ell(\theta^*, X)]^{-1})_{jj}).$$

The assumption that the Fisher information $\mathbb{E}[\nabla_\theta^2 \ell(\theta^*, X)]^{-1}$ is sparse is very hard to check in practice (e.g., in logistic regression); an open problem in high-dimensional inference is whether it is possible to derive the asymptotic normality of general regularized M-estimator without such sparsity assumption.

Chapter 19
Multiple Hypotheses

19.1 Conformal Inference

The conformal inference aims to build confidence intervals for predictions without any unnecessary assumptions, especially those about models. Given i.i.d. random pairs $(X_i, Y_i) \sim P$, our goal is to build a confidence interval $\widehat{C}_\alpha$ for the prediction given the new pair (X_{n+1}, Y_{n+1}), which has

$$\mathbb{P}(Y_{n+1} \in \widehat{C}_\alpha(X_{n+1})) \geq 1 - \alpha.$$

In order to explain the idea of conformal inference, we start with Y_i's themselves. Denote

$$\widehat{q}_n = \text{Quantile}(1 - \alpha, \{Y_i\}_{i=1}^n)$$

$$\mathbb{P}(Y_{n+1} \leq \widehat{q}_n) \approx 1 - \alpha.$$

By symmetry, the rank of Y_{n+1} among $Y_1, \ldots, Y_{n+1}$ follows a uniform distribution $\text{Unif}\{1, 2, \ldots, n+1\}$. We could construct

$$\pi(y) = \frac{1}{n+1} \sum_{i=1}^{n} 1\{Y_i \leq y\} + \frac{1}{n+1}.$$

And by the symmetry, we have

$$\mathbb{P}\left(\pi(Y_{n+1}) \leq \frac{\lceil (n+1)(1-\alpha) \rceil}{n+1}\right) \geq 1 - \alpha.$$

We can invert the $\pi(\cdot)$ above and solve Y_{n+1} as

J. Lu, *Big Data Analysis*, https://doi.org/10.1007/978-3-032-03161-7_19

$$\mathbb{P}(Y_{n+1} \leq \widetilde{q}_n) \geq 1-\alpha, \text{ where } \widetilde{q}_n = \text{Quantile}(1-\alpha, \{Y_i\}_{i=1}^n \bigcup \{\infty\}).$$

However, the above confidence interval has not involved X_{n+1}. Let $\widehat{f}(x)$ be some estimator of $\mathbb{E}[Y|X=x]$ using the data $\{(X_i, Y_i)\}_{i=1}^n$.

Denote the absolute residual $R_i = |Y_i - \widehat{f}(X_i)|$, and we may use a similar idea as above by considering

$$\widetilde{q}_n = \text{Quantile}(1-\alpha, \{R_i\}_{i=1}^n \bigcup \{\infty\}).$$

One possible confidence interval could be

$$\widehat{C}_n(x) = [\widehat{f}(x) - \widehat{q}_n, \widehat{f}(x) + \widehat{q}_n].$$

However, such confidence interval is not honest. The reason is that R_i depends on $\widehat{f}$, which further depends on $\{(X_i, Y_i)\}_{i=1}^n$. As a result, R_i could not be independent and identically distributed. To handle this issue, we would estimate $\widehat{f}_{n,(x,y)}$ using the data

$$\{(X_1, Y_{n+1}), (X_2, Y_{n+1}), \ldots, (X_n, Y_{n+1}), (x, y)\}.$$

Under this framework, we define the new residuals

$$\begin{aligned} R_i^{(x,y)} &= |Y_i - \widehat{f}_{n,(x,y)}(X_i)| \text{ if } i = 1, 2, \ldots, n, \\ R_{n+1}^{(x,y)} &= |y - \widehat{f}_{n,(x,y)}(x)|. \end{aligned}$$

By symmetry, now we have the rank of $R_{n+1}^{(x,y)}$ among $R_i^{(x,y)}, i \in \{1, 2, \ldots, n+1\}$ follow Unif$(1, 2, \ldots, n+1)$. So we can construct the confidence interval via

$$\begin{aligned} \widetilde{q}_n &= \text{Quantile}(1-\alpha, \{R_i^{(x,y)}\}_{i=1}^n \bigcup \{\infty\}) \\ \widehat{C}_n(x) &= [\widehat{f}(x) - \widehat{q}_n, \widehat{f}(x) + \widehat{q}_n] \end{aligned}$$

We can show that

$$\mathbb{P}(Y_{n+1} \in \widehat{C}_n(X_{n+1})) \geq 1-\alpha.$$

The reason why we do so is that (1) we would not need any unnecessary assumptions, (2) we would not need to use any asymptotic properties, (3) we could avoid overfitting since for most of the time the naive CI would undercover, and (4) such R_i could be generalized in several more sophisticated frameworks.

19.2 Multiple Hypotheses

Here we are discussing the hypotheses $\{H_{0i}\}_{i=1}^N$, with

$$H_{0i} : \theta_i^* = \theta_{i0}; \quad H_{1i} : \theta_i^* \neq \theta_{i0}.$$

In some practical scenarios the number of assumptions, N could be very large. N may be on the level of several millions on GWAS when our goal is to detect significant genes. We would hope to find a way to control the family-wise error rate (FWER), which could be written as

$$\text{FWER} = \mathbb{P}(\text{There is at least one type-I error}) \leq \alpha.$$

19.2.1 P-Values

For a single null hypothesis, we could either construct a statistic T and a corresponding confidence region such that the rejection region R_α holds that

$$P_0(T \in R_\alpha) = \alpha.$$

We can define the p-value as

$$p(T) = \inf_\alpha P_0(T \in R_\alpha).$$

Since $p(T) \sim \text{Unif}[0, 1]$ under the null hypothesis, we will tend to reject the null hypothesis H_0 if $p(T)$ is smaller than the significance level α.

19.2.2 Bonferroni Correction

Here we use the notations given above and suppose that we have p_i as the p-value of a null hypothesis H_{i0}. For a significance level α, the **Bonferroni correction** rejects H_{i0} if $p_i \leq \frac{\alpha}{N}$. In this case, we would have

$$\begin{aligned}\text{FWER} &= \mathbb{P}(\text{There is at least one type-I error}) \\ &\leq \mathbb{P}\left(\bigcup_{i=1}^N \{\psi_i = 1, i \in H_0\}\right) \leq \sum_{i=1}^N \mathbb{P}(\psi_i = 1, i \in H_0) = \sum_{i=1}^N \frac{\alpha}{N} = \alpha.\end{aligned}$$

The Bonferroni is usually **too conservative**, as the step we use union bound could make the inequality very loose. In fact, we have not utilized the dependency of p_i in Bonferroni correction.

For example, we could consider a debiased Lasso model $Y = \mathbb{X}\beta^* + \epsilon$ and have $H_{0j} : \beta_j^* = 0$ as our multiple hypotheses. According to the previous chapter, we have that given some regularization conditions, the debiased Lasso estimator $\widehat{\beta}^d$ would have

$$\sqrt{n}(\widehat{\beta}_j^d - \beta_j^*) \xrightarrow{d} N(0, \Theta_{jj})$$

for any arbitrary j and Θ is the precision matrix. If we consider Bonferroni correction, we could calculate p_j as

$$p_j = 1 - 2\Phi\left(\frac{\sqrt{n}|\widehat{\beta}_j^d|}{\sqrt{\widehat{\Theta}_{jj}}}\right)$$

and reject H_{0j} if $p_j \leq \frac{\alpha}{d}$. This will be conservative when the dimension d is large.

19.2.3 Maximal Statistic

An alternative to Bonferroni correction is that we calculate the maximal statistic as

$$T = \max_{1 \leq j \leq d} T_j = \max_{1 \leq j \leq d} \frac{\sqrt{n}|\widehat{\beta}_j^d|}{\sqrt{\widehat{\Theta}_{jj}}}$$

If we can estimate the quantile of the maximal statistic $\widehat{C}(1-\alpha)$ such that

$$\mathbb{P}(T > \widehat{C}(1-\alpha)) = \alpha$$

we would reject H_{0j} if $T_j > \widehat{C}(1-\alpha))$. Then the FWER becomes

$$\begin{aligned}
\text{FWER} &= \mathbb{P}(\text{There is at least one type-I error}) \\
&\leq \mathbb{P}\left(\bigcup_{j=1}^{d} \{T_j > \widehat{C}(1-\alpha)\}\right) \\
&= 1 - \mathbb{P}(T_j \leq \widehat{C}(1-\alpha) \text{ for any } j) = 1 - \mathbb{P}(\max_j T_j \leq \widehat{C}(1-\alpha)) = \alpha.
\end{aligned}$$

Now we could save much p-values by using the maximal statistic compared to Bonferroni. In the next chapter, we are going to discuss how to estimate the quantile of maximal statistics.

Chapter 20
False Discovery Rate

20.1 Gaussian Multiplier Bootstrap

We will continue the discussion on control family-wise error rate via the maximal statistic. Given hypotheses $\{H_{0i}\}_{i=1}^{N}$, for each H_{0j}, we have a statistic T_j such that for a single hypothesis, we will reject H_{0j} if $T_j \geq q_\alpha$ where $q_\alpha = \arg\min_t \mathbb{P}_{H_0}(|T_j| > t) \leq \alpha$. In the previous chapter, we introduced the maximal statistic $T = \max_j |T_j|$. If we can estimate the quantile of T:

$$\widehat{C}(1-\alpha) = \text{Quantile}(\max_j |T_j|, 1-\alpha), \tag{20.1}$$

we reject H_{0j} if $T_j \geq \widehat{C}(1-\alpha)$ for all $1 \leq j \leq N$. We showed that such maximal statistic approach obtains FWER $< \alpha$. Here, we consider how to estimate the $(1-\alpha) \times 100\%$ quantile of the $\max_j T_j$. Let us start with a simple case: the statistics $T_1, \ldots, T_N$ are i.i.d. $N(0, 1)$, then we solve the following equation:

$$\begin{aligned}\mathbb{P}\big(\max_j |T_j| > t\big) &= \mathbb{P}\Big(\bigcup_{j=1}^{d}\{|T_j| > t\}\Big) = 1 - \mathbb{P}\Big(\bigcap_{j=1}^{d}\{|T_j| \leq t\}\Big) \\ &= 1 - \prod_{j=1}^{d}[1 - 2\Phi(1-t)] = \alpha,\end{aligned}$$

where Φ denotes the cdf of the standard normal distribution. So the $(1-\alpha) \times 100\%$ quantile of the $T = \max_j T_j$ is

$$\widehat{C}(1-\alpha) = 1 - \Phi^{-1}\big([1-(1-\alpha)^{1/d}]/2\big).$$

J. Lu, *Big Data Analysis*, https://doi.org/10.1007/978-3-032-03161-7_20

The case where the T_j are dependent and only asymptotically normal is more challenging. Typically, such T_j can be decomposed as

$$T_j = \frac{1}{\sqrt{n}} \sum_{i=1}^{n} \big(f_j(X_i) - \mathbb{E}[f_j(X)]\big) + o_P(1), \tag{20.2}$$

where $f_j : \mathbb{R}^d \to \mathbb{R}$ is some deterministic function and the first term on the right-hand side is the leading term contributing to asymptotic normality of the statistic. For example, in Chap. 18, we showed that the debiased Lasso has the decomposition

$$\sqrt{n}(\widehat{\beta}_j^d - \beta_j^*) = \frac{1}{\sqrt{n}} \sum_{i=1}^{n} \Theta_j^\top X_i \varepsilon_i + o_P(1).$$

So we can conduct the hypotheses $H_{0j} : \beta_j^* = 0$ by the statistic $\sqrt{n}|\widehat{\beta}_j^d|$.[1] In order to estimate the quantile of $\max_j T_j$, our idea is to estimate the quantile of the maximal of the leading term:

$$T_0 = \max_j \frac{1}{\sqrt{n}} \sum_{i=1}^{n} \big(f_j(X_i) - \mathbb{E}[f_j(X)]\big).$$

Denote $f(X) = (f_1(X), \ldots, f_d(X))^\top \in \mathbb{R}^d$. By central limit theorem, we have

$$\frac{1}{\sqrt{n}} \sum_{i=1}^{n} \big(f(X_i) - \mathbb{E}[f(X)]\big) \xrightarrow{d} N(0, \mathrm{Cov}(f(X))).$$

Let $Z \sim N(0, \mathrm{Cov}(f(X)))$, and we will expect that if a random vector converges in distribution to a normal vector, then the distribution of their maximal statistics should be close as well, i.e.,

$$T_0 = \max_j \frac{1}{\sqrt{n}} \sum_{i=1}^{n} \big(f_j(X_i) - \mathbb{E}[f_j(X)]\big) \overset{d}{\approx} \max_j Z_j.$$

So the quantile $\widehat{C}(1-\alpha)$ in (20.1) has

$$\widehat{C}(1-\alpha) \approx \mathrm{Quantile}(\max_j |T_{0j}|, 1-\alpha) \approx \mathrm{Quantile}(\max_j |Z_j|, 1-\alpha), \tag{20.3}$$

[1] We do not standardize the statistic only for notation simplicity. The standardized t-statistic $\sqrt{n}\widehat{\Theta}_{jj}^{-1/2}|\widehat{\beta}_j^d|$ also has the decomposition (20.2).

where we can estimate the last term as long as we can sample the multivariate normal $N(0, \mathrm{Cov}(f(X)))$. However, in most cases, we do not know the covariance. For example, in the debiased Lasso,

$$f(X, \varepsilon) = \Theta X\varepsilon \text{ and } \mathrm{Cov}(f(X, \varepsilon)) = \sigma^2\Theta,$$

where $\mathrm{Var}(\varepsilon) = \sigma^2$. One solution is to plug in the CLIME estimator $\widehat{\Theta}$ and residual variance estimator $\widehat{\sigma}^2 = \frac{1}{n}\sum_{i=1}^{2}(Y_i - X_i\widehat{\beta})^2$, where $\widehat{\beta}$ is the Lasso estimator. We can simulate $\widetilde{Z} \sim N(0, \widehat{\sigma}^2\widehat{\Theta})$ and estimate the quantile $\widehat{C}(1-\alpha) \approx \max_j \mathrm{Quantile}(\max_j |\widetilde{Z}_j|, 1-\alpha)$. However, it is not always possible to have a closed form for the covariance $\mathrm{Cov}(f(X))$. We will provide a more general approach to estimate the quantile of T_0 via the so-called Gaussian multiplier bootstrap.

Let $\xi_1, \ldots, \xi_n$ be i.i.d. samples of $N(0, 1)$; we define the **Gaussian multiplier bootstrap** statistic as

$$T_\xi = \frac{1}{\sqrt{n}}\sum_{i=1}^{n}\big(f(X_i) - \mathbb{E}[f(X)]\big)\xi_i.$$

As $T_\xi \big| \{X_i\}_{i=1}^n \sim N(0, \widehat{\mathrm{Cov}}(f(X)))$, where the sample covariance

$$\widehat{\mathrm{Cov}}(f(X)) = \frac{1}{n}\sum_{i=1}^{n}\big(f(X_i) - \mathbb{E}[f(X)]\big)\big(f(X_i) - \mathbb{E}[f(X)]\big)^\top \approx \mathrm{Cov}(f(X)).$$

We will expect for the multivariate normal $Z \sim N(0, \mathrm{Cov}(f(X)))$, there is

$$\arg\min_t\{\mathbb{P}(\max_j |T_{\xi,j}| \geq t | \{X_i\}_{i=1}^n) \leq \alpha\} \approx \mathrm{Quantile}(\max_j |Z_j|, 1-\alpha).$$

Combining with (20.3), we estimate the quantile of T_0 via

$$\widehat{C}(1-\alpha) \approx \arg\min_t\{\mathbb{P}(\max_j |T_{\xi,j}| \geq t | \{X_i\}_{i=1}^n) \leq \alpha\}.$$

However, recall that for the debiased Lasso, $f(X, \varepsilon) = \Theta X\varepsilon$, but we do not know Θ and ε. The following part summarizes the procedure to estimate the quantile when $f(X)$ is unknown.

Let $\xi_1, \ldots, \xi_n$ be i.i.d. samples of $N(0, 1)$ and $\widehat{f}(X)$ is a consistent estimate of $f(X) - \mathbb{E}[f(X)]$. We construct the statistic

(continued)

$$\widehat{T}_\xi = \frac{1}{\sqrt{n}}\sum_{i=1}^{n}\widehat{f}(X_i)\xi_i.$$

The quantile of T_0 can be estimated as

$$\widehat{C}(1-\alpha) \approx \arg\min_t\{\mathbb{P}(\max_j |\widehat{T}_{\xi,j}| \geq t|\{X_i\}_{i=1}^n) \leq \alpha\}. \tag{20.4}$$

For the debiased Lasso, we can estimate $f(X_i, \varepsilon_i) = \Theta X_i \varepsilon_i$ via $\widehat{f}(X_i, \varepsilon_i) = \widehat{\Theta} X_i \widehat{\varepsilon}_i$, where $\widehat{\Theta}$ is the CLIME estimator and $\widehat{\varepsilon}_i = Y_i - X_i\widehat{\beta}$. We construct the multiplier Gaussian bootstrap statistic

$$\widehat{T}_\xi = \frac{1}{\sqrt{n}}\sum_{i=1}^{n}\widehat{\Theta} X_i \widehat{\varepsilon}_i \xi_i.$$

Let $\widehat{C}(1-\alpha)$ be the estimator of the quantile of $\max_j \sqrt{n}(\widehat{\beta}_j^d - \beta_j^*)$ via (20.4). We can reject $H_{0j} : \beta_j^* = 0$ if $\sqrt{n}|\widehat{\beta}_j^d| \geq \widehat{C}(1-\alpha)$ for all $1 \leq j \leq d$ such that FWER $\leq \alpha$.

20.2 False Discovery Rate: Independent P-Values

As we have commented in Lecture 16, the family-wise error rate might be too stringent that we are not allowed to make any type-I errors and we define the false discovery rate to make our inference more liberal.

Let $\psi_i = 1$ if we reject H_{0i} and $\psi_i = 0$ otherwise for $1 \leq i \leq N$. Recall that the false discovery proportion (FDP) and the false discovery rate are

$$\text{FDP} = \frac{\#\text{False Positive}}{\#\text{Rejected Hypotheses}} = \frac{\sum_{i\in H_0}\psi_i}{\max}\{\sum_{i=1}^{d}\psi_i, 1\} \text{ and FDR} = \mathbb{E}[\text{FDP}].$$

We aim to find such tests $\{\psi_i\}_{i=1}^N$ such that FDR $= \mathbb{E}[\text{FDP}] \leq \alpha$. Let p_i be the p-value H_{0i} and we first consider the case the p-values $p_1, \ldots, p_N$ are independent. We have the following procedure to control the FDR.

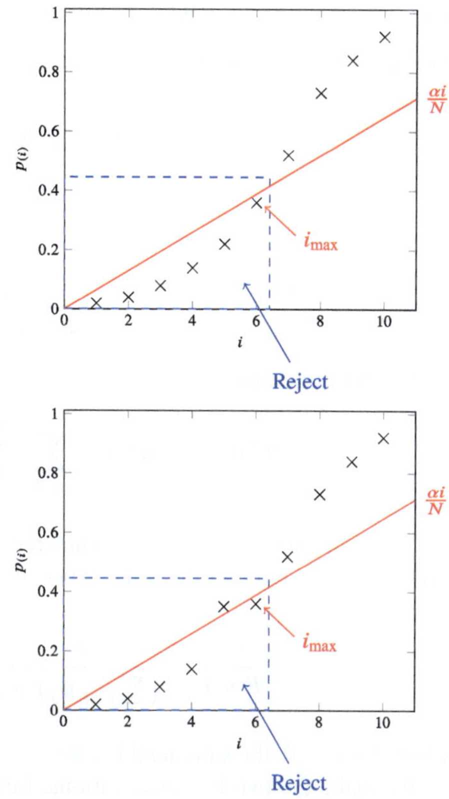

Fig. 20.1 Two examples of the Benjamini-Hochberg procedure. The red line illustrates the threshold for $p_{(i)}$. The blue rectangle illustrates the region rejection

> We order the p-values as $p_{(1)} \leq \cdots \leq p_{(N)}$. Let $i_{\max} = \arg\max\{i : p_{(i)} \leq \frac{i\alpha}{N}\}$. The **Benjamini-Hochberg** (BH_α) procedure rejects $H_{0(i)}$ if $i \leq i_{\max}$.

Two examples of the BH_α procedure are given in Fig. 20.1. In the example in the right panel of Fig. 20.1, notice that there is an i such that $p_{(i)} > \frac{\alpha i}{N}$ but $i < i_{\max}$. Therefore, we would still reject this H_{0i} based on this procedure. Notice that the BH_α is more liberal than the Bonferroni correction as there is an additional i in the α/N term.

The following theorem shows that Benjamini-Hochberg procedure can control FDR for independent p-values.

Theorem 20.1 *Suppose that $p_1, \ldots, p_N$ are independent. Then BH_α controls the FDR as*

$$\mathrm{FDR} = \mathbb{E}[\mathrm{FDP}_{\mathrm{BH}_\alpha}] = \frac{N_0}{N}\alpha \leq \alpha,$$

where N_0 denotes the number of true H_{0i}'s.

Proof Denote the number of rejections $R = \sum_{j=1}^{N} \psi_j$, then

$$\text{FDP} = \sum_{i \in H_0} \frac{\psi_i}{R \vee 1}.$$

We claim that

$$\mathbb{E}[\frac{\psi_i}{R \vee 1}] = \frac{\alpha}{N}, \tag{20.5}$$

as it then follows that

$$\text{FDR} = \mathbb{E}[\text{FDP}] = \sum_{i \in H_0} \mathbb{E}\Big[\frac{\psi_i}{R \vee 1}\Big] = \frac{N_0 \alpha}{N} \le \alpha,$$

which completes the proof. So what remains is the prove the claim (20.5). We decompose the ratio inside of the expectation as a summation

$$\frac{\psi_i}{R \vee 1} = \frac{\psi_i}{\big(\sum_{j=1}^{N} \psi_j\big) \vee 1} = \sum_{k=1}^{N} \frac{\psi_i \mathbb{1}(R = k)}{k \vee 1},$$

where for $k = 0$, the summand is zero.

We make the two key observations. First, when $R = k$, $\psi_i = 1$ if and only if $p_i \le \frac{\alpha k}{N}$ because $p_{(k)} \le \frac{\alpha k}{N}$ and we reject H_{0i} if $p_i \le p_{(k)}$.

A challenge in evaluating $\mathbb{E}[\psi_i \mathbb{1}(R = k)]$ is that R and ψ_i are not independent. Consider the following leave-one-out approach. If $p_i \le \frac{\alpha k}{N}$, then we set p_i to 0, denoted by $p_i \to 0$. Let $R(p_i \to 0)$ denote the number of rejected hypotheses by BH_α using the p-values: $p_1, \ldots, p_{i-1}, 0, p_{i+1}, \ldots, p_N$. From Fig. 20.2, our second key observation is that if $p_i \le \frac{\alpha k}{N}$, then we set p_i to 0 and the number of rejections is the same as R. Namely, $\psi_i \mathbb{1}(R = k) = \psi_i \mathbb{1}(R(p_i \to 0) = k)$.

Let $\mathcal{F}_i = \{p_1, p_2, \ldots, p_{i-1}, p_{i+1}, \ldots, p_N\}$. Then, using the two observations above, we have

$$\begin{aligned}
\mathbb{E}\Big[\frac{\psi_i}{R \vee 1}\Big|\mathcal{F}_i\Big] &= \mathbb{E}\Big[\sum_{k=1}^{N} \frac{\psi_i \mathbb{1}(R = k)}{k \vee 1}\Big|\mathcal{F}_i\Big] \\
&= \sum_{k=1}^{N} \mathbb{E}\Big[\frac{\psi_i \mathbb{1}(R = k)}{k}\Big|\mathcal{F}_i\Big] \\
&= \sum_{k=1}^{N} \mathbb{E}\Big[\frac{\mathbb{1}(p_i \le \frac{\alpha k}{N}) \mathbb{1}(R(p_i \to 0) = k)}{k}\Big|\mathcal{F}_i\Big]
\end{aligned}$$

Fig. 20.2 Two examples of the Benjamini-Hochberg procedure. The panel in the right illustrates how setting $p_{(5)}$ to 0 does not affect the rejection region

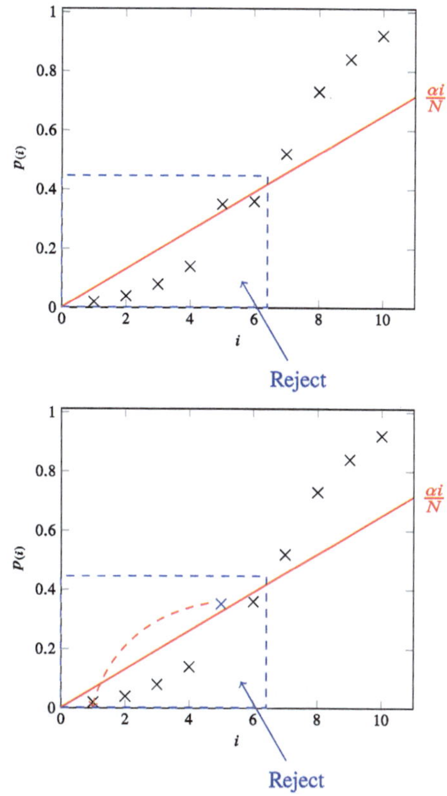

$$= \sum_{k=1}^{N} \frac{\frac{\alpha k}{N} \mathbb{1}(R(p_i \to 0) = k)}{k} = \frac{\alpha}{N} \sum_{k=1}^{N} \mathbb{1}(R(p_i \to 0) = k) = \frac{\alpha}{N},$$

where in the third equality, we use the two key observations and in the fourth equality we use $R(p_i \to 0)$ is irrelevant to p_i. Therefore, by tower property, we have

$$\mathbb{E}\Big[\frac{\psi_i}{R \vee 1}\Big] = \mathbb{E}\Big[\mathbb{E}\Big[\frac{\psi_i}{R \vee 1}\Big|\mathcal{F}_i\Big]\Big] = \frac{\alpha}{N}$$

and we prove the claim (20.5). □

Fig. 20.2 Two examples of the Benjamini–Hochberg procedure. The panel in the right illustrates how setting $p_{(i)}$ to 0 does not affect the rejection region

Chapter 21
Knock-Off

21.1 False Discovery Rate: Dependent P-Values

We will continue the discussion of controlling the false discovery rate (FDR). When testing null hypotheses $\{H_{0j}\}_{j=1}^d$, recall that the FDR is defined as

$$\text{FDR} = \mathbb{E}\left(\frac{\#\text{False Positives}}{\#\text{Rejected Hypotheses}}\right).$$

In the previous chapter, we discussed the case where the p-values corresponding to the $\{H_{0j}\}_{j=1}^d$ were independent. Here, we consider the more challenging case where the p-values are dependent.

Throughout, we consider the following framework. Given a response Y (e.g., phenotype) and features $X_1, \ldots, X_d$ (e.g., SNPs), our goal is to select features X_j's related to Y. Since the X_j's may be correlated, we consider testing the null hypothesis $Y \perp\!\!\!\perp X_j | X_{-j}$. In the example of the linear model $Y = \mathbb{X}\beta + \epsilon$ where $\mathbb{X} \in \mathbb{R}^{n\times d}$, we are interested in testing the null hypothesis $\beta_j = 0$.

21.2 Permutation Test

One popular approach for controlling the FDR in this context is the **permutation test**. Consider the Lasso estimator which we denote as $\widehat{\beta} \leftarrow \text{Lasso}(\mathbb{X}, Y)$. Let $\mathbb{X}_\pi$ denote the matrix obtained by permuting the rows of $\mathbb{X}$ and consider the Lasso estimator based on $\mathbb{X}_\pi$, denoted by $\widetilde{\beta} \leftarrow \text{Lasso}(\mathbb{X}_\pi, Y)$.

The idea behind the permutation test is that we expect $\widehat{\beta} \overset{d}{=} \widetilde{\beta}$. However, one must be careful because this is not always true, which can be illustrated in the following counterexample. Suppose that $Y = X_1 + \epsilon$ where $\mathbb{E}(X_1) = \mathbb{E}(X_2) = 0$, $\text{Var}(X_1) = \text{Var}(X_2) = 1$, and $\text{Corr}(X_1, X_2) = 1/2$. A straightforward calculation

J. Lu, *Big Data Analysis*, https://doi.org/10.1007/978-3-032-03161-7_21

shows that $\mathbb{E}(YX_1) = 1$ and $\mathbb{E}(YX_2) = 0.5$. Letting $X_{2,\pi}$ denote the vector obtained by permuting the entries of X_2, it can be shown that $\mathbb{E}(YX_{2,\pi}) = 0$ which implies $YX_2 \overset{d}{\neq} YX_{2,\pi}$.

21.3 Knock-Off

As the permutation test is not always valid, we may instead consider the **knock-off** approach for controlling the FDR, which is a new and active area of research. Suppose that we have a test statistic Z_j that depends on $\mathbb{X}$ and Y, such as the t-statistic of β_j in Lasso. The idea behind this approach is that we construct dummy variables $\widetilde{\mathbb{X}}$ such that $Z_j(\mathbb{X}, Y) \overset{d}{=} Z_j(\widetilde{\mathbb{X}}, Y)$.

Suppose that we wish to select from d candidate features. Denote

$$[\underbrace{Z_1, \ldots, Z_d}_{\text{Original}}, \underbrace{\widetilde{Z}_1, \ldots, \widetilde{Z}_d}_{\text{Knock-Off}}] = Z([\mathbb{X}, \widetilde{\mathbb{X}}], Y).$$

As a prototypical example, we consider a linear model where the outcome is Y and the design matrix is $[\mathbb{X}, \widetilde{\mathbb{X}}]$, as illustrated in Fig. 21.1.

If $j \in H_{0j}$, then we expect that $(Z_j, \widetilde{Z}_j) \overset{d}{=} (\widetilde{Z}_j, Z_j)$ (i.e., exchangeability holds). This notion can be generalized as follows. Let $(\widetilde{Z}, Z)_{\text{swap}(T)}$ denote the vector obtained by swapping the set $T \subset \{1, \ldots, d\}$ of the entries of Z with the corresponding entries of $\widetilde{Z}$ in $(Z, \widetilde{Z})$, as depicted in Fig. 21.2. We expect that for any $T \subset H_0$, $(Z, \widetilde{Z}) \overset{d}{=} (\widetilde{Z}, Z)_{\text{swap}(T)}$. In order to find an $\widetilde{\mathbb{X}}$ such that $(Z, \widetilde{Z}) \overset{d}{=} (\widetilde{Z}, Z)_{\text{swap}(T)}$ is satisfied, we note that $\mathbb{X} \overset{d}{=} \widetilde{\mathbb{X}}$ implies $(Z, \widetilde{Z}) \overset{d}{=} (\widetilde{Z}, Z)_{\text{swap}(T)}$.

Definition 21.1 (Knock-Off Score) The knock-off score is a statistic $W_j = w_j(Z_j, \widetilde{Z}_j)$ such that

1. The w_j are antisymmetric, i.e., $w_j(x, y) = -w_j(y, x)$ for $1 \leq j \leq d$.
2. If $j \in H_0$, then $W_j \overset{d}{=} -\widetilde{W}_j$.

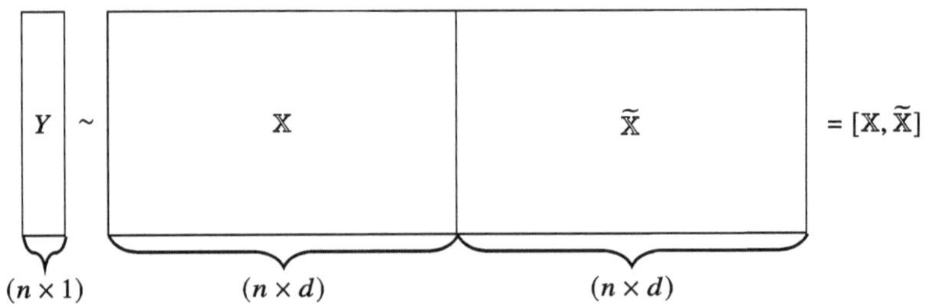

Fig. 21.1 A linear model where the outcome is Y and design matrix is $[\mathbb{X}, \widetilde{\mathbb{X}}]$

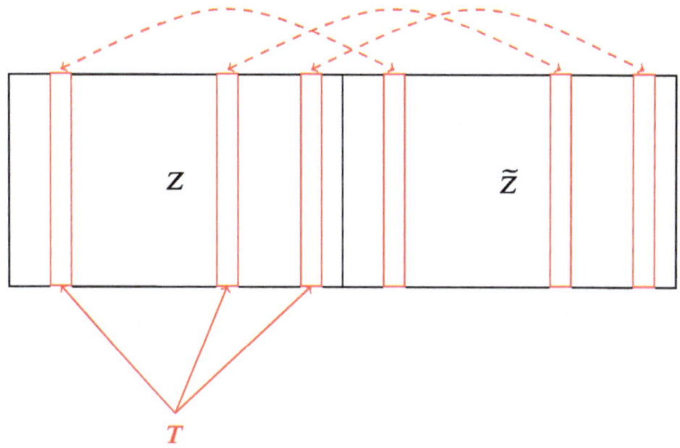

Fig. 21.2 The construction of $(\widetilde{Z}, Z)_{\text{swap}(T)}$, where $Z, \widetilde{Z} \in \mathbb{R}^d$ and $T \subset \{1, \ldots, d\}$

3. The $\text{sign}(W_j) || W_1|, \ldots, |W_d|$ are i.i.d. Ber(1/2).

For instance, it can be easily verified that $W_j = Z_j - \widetilde{Z}_j$ is a valid knock-off score if $(Z_j, \widetilde{Z}_j) \overset{d}{=} (\widetilde{Z}_j, Z_j)$.

Consider a procedure where we reject H_{0j} if $W_j \geq t$ for some threshold level t. By definition,

$$\text{FDP}(t) = \frac{\#\{j \in H_0 : W_j \geq t\}}{\#\{j : W_j \geq t\} \vee 1}.$$

To estimate FDP(t), it follows from Condition 2 in Definition 21.1 that

$$\text{FDP}(t) \approx \frac{\#\{j \in H_0 : W_j \leq -t\}}{\#\{j : W_j \geq t\} \vee 1} \leq \underbrace{\frac{\#\{j : W_j \leq -t\} + 1}{\#\{j : W_j \geq t\} \vee 1}}_{\widehat{\text{FDP}}(t)}.$$

Letting

$$S^+(t) = \{j : |W_j| \geq t, \text{sign}(W_j) = 1\}$$
$$S^-(t) = \{j : |W_j| \geq t, \text{sign}(W_j) = -1\},$$

we can express $\widehat{\text{FDP}}(t)$ more simply by

$$\widehat{\text{FDP}}(t) = \frac{\#S^-(t) + 1}{\#S^+(t) \vee 1}.$$

The knock-off procedure (illustrated in Fig. 21.3) can then be described as follows:

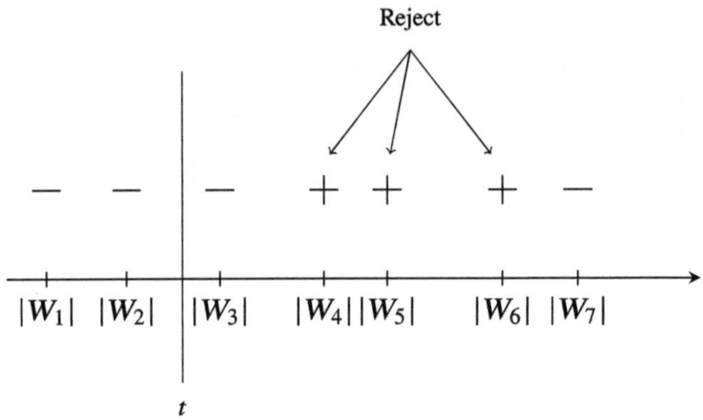

Fig. 21.3 Schematic of the knock-off procedure. The line indexed by t illustrates the threshold level. The sign above $|W_j|$ indicates the sign of W_j. According to the procedure, we reject null hypotheses corresponding to the test statistics W_4, W_5, and W_6

> Let $t^* = \arg\min_t\{\widehat{\mathrm{FDP}}(t) \le \alpha\}$. The knock-off procedure rejects H_{0j} if and only if $j \in S^+(t^*)$.

The validity of the knock-off procedure is proven in Theorem 21.2, which requires some results on martingales that can be summarized as follows. A sequence of random variables $\{X_i\}_{i=1}^{\infty}$ is called a martingale if $\mathbb{E}(X_{n+1}|X_1, \ldots, X_n) = X_n$ almost surely for all $n \ge 1$ and is called a supermartingale if $\mathbb{E}(X_{n+1}|X_1, \ldots, X_n) \le X_n$ almost surely for all $n \ge 1$. For instance, if $\xi_i = \pm 1$ with probability 1/2, then $X_n = \sum_{i=1}^{n} \xi_i$ is a martingale. In the context of a gambling system, one's income is a martingale in a fair game and is a supermartingale in an unfair game. Lastly, we will need the following result on martingales and supermartingales in the proof of Theorem 21.2.

Theorem 21.1 (Optimal Stopping Theorem) *Suppose that the stopping time* τ *only depends on the current information. If* X_t *is a martingale, then* $\mathbb{E}(X_\tau) = \mathbb{E}(X_0)$. *If* X_t *is supermartingale, then* $\mathbb{E}(X_\tau) \le \mathbb{E}(X_0)$.

Theorem 21.2 (Knock-Off) *Suppose that we reject* H_{0j} *if and only if* $j \in S^+(t^*)$ *where* $t^* = \arg\min_t\{\widehat{FDP}(t) \le \alpha\}$. *Then,*

$$\mathbb{E}\left(\frac{\#False\ positives}{\#Rejected\ Hypotheses \vee 1}\right) = \mathbb{E}(\widehat{FDP}(t)) \le \alpha.$$

Proof We have that

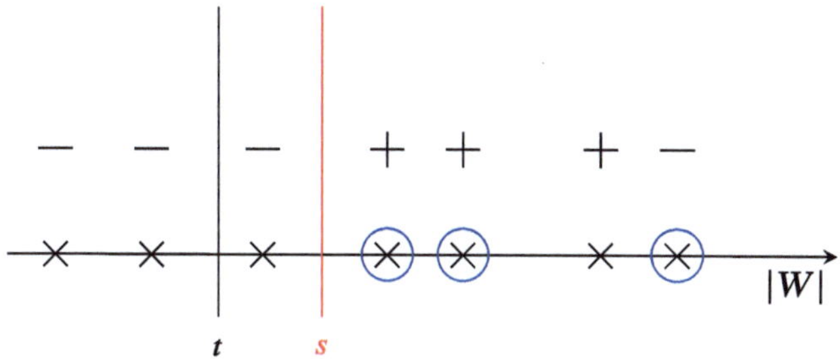

Fig. 21.4 Schematic of the knock-off procedure. The lines indexed by t and s illustrate two threshold levels. The sign above $|W_j|$ indicates the sign of W_j. The blue circles indicate the null hypotheses that are true

$$
\begin{aligned}
\mathrm{FDP}(t^*) &= \frac{\#\{j \in H_0 : j \in S^+(t^*)\}}{\#\{j : j \in S^+(t^*)\}} \left(\frac{1 + \#\{j \in H_0 : j \in S^-(t^*)\}}{1 + \#\{j \in H_0 : j \in S^-(t^*)\}} \right) \\
&\leq \widehat{\mathrm{FDP}}(t^*) \frac{\#\{j \in H_0 : j \in S^+(t^*)\}}{1 + \#\{j \in H_0 : j \in S^-(t^*)\}} \\
&\leq \alpha \frac{V_+(t^*)}{1 + V_-(t^*)}
\end{aligned}
$$

where $V_\pm(t^*) = \#\{j \in H_0 : j \in S^\pm(t^*)\}$.

It suffices to show that $\mathbb{E}\left(\frac{V_+(t^*)}{1+V_-(t^*)}\right) \leq 1$. A key observation is that $V^+(t)|V^-(t)+V^+(t)$ follows a hypergeometric distribution, as the sign$(W_j)|W$ are i.i.d. Ber$(1/2)$. Our goal then is to show that for $s \geq t$,

$$
\mathbb{E}\left(\frac{V_+(s)}{1 + V_-(s)} | \mathcal{F}_t \right) \leq \frac{V_+(t)}{1 + V_-(t)}
$$

where $\mathcal{F}_t$ contains information if $j \in H_0$ for $|W_j| > t$ and $V_-(t) + V_+(t)$. See Fig. 21.4 for an illustration. Since $\frac{V_+(t^*)}{1+V_-(t^*)}$ is supermartingale, it follows from Theorem 21.1 that

$$
\mathbb{E}\left(\frac{V_+(t^*)}{1 + V_-(t^*)} \right) \leq \mathbb{E}\left(\frac{V_+(0)}{1 + V_-(0)} \right) \leq 1
$$

which completes the proof. □

References

1. R. Adamczak, A tail inequality for suprema of unbounded empirical processes with applications to Markov chains. Electron. J. Probab. **34**, 1000–1034 (2008)
2. A. Belloni, V. Chernozhukov, L. Wang, Square-root lasso: pivotal recovery of sparse signals via conic programming. Biometrika **98**(4), 791–806 (2011)
3. T. Cai, W. Liu, A direct estimation approach to sparse linear discriminant analysis. J. the Am. Statist. Assoc. **106**(496), 1566–1577 (2011)
4. V. Chernozhukov, D. Chetverikov, K. Kato, Gaussian approximations and multiplier bootstrap for maxima of sums of high-dimensional random vectors. Ann. Stat. **41**, 2786–2819 (2013)
5. J. Chiquet, Y. Grandvalet, C. Charbonnier, Sparsity with sign-coherent groups of variables via the cooperative-lasso. Ann. Appl. Stat. **6**(2), 795–830 (2012)
6. M.D. Escobar, M. West, Bayesian density estimation and inference using mixtures. J. Am. Stat. Assoc. **90**(430), 577–588 (1995)
7. J. Fan, R. Li, Variable selection via nonconcave penalized likelihood and its oracle properties. J. Am. Stat. Assoc. **96**(456), 1348–1360 (2001)
8. J. Fan, R. Li, C.-H. Zhang, H. Zou, *Statistical Foundations of Data Science* (CRC Press, Boca Raton, 2020)
9. J. Friedman, T. Hastie, R. Tibshirani, Sparse inverse covariance estimation with the graphical lasso. Biostatistics **9**(3), 432–441 (2008)
10. I.K. Glad, N.L. Hjort, N.G. Ushakov, Correction of density estimators that are not densities. Scand. J. Stat. **30**, 415–427 (2003)
11. H. Ishwaran, L.F. James, Approximate Dirichlet process computing in finite normal mixtures: smoothing and prior information. J. Comput. Graph. Stat. **11**(3), 508–532 (2002)
12. P.-L. Loh, M.J. Wainwright, Regularized M-estimators with nonconvexity: statistical and algorithmic theory for local optima. J. Mach. Learn. Res. **16**(1), 559–616 (2015)
13. N. Meinshausen, Relaxed lasso. Comput. Stat. Data Anal. **52**(1), 374–393 (2007)
14. P. Ravikumar, J. Lafferty, H. Liu, L. Wasserman, Sparse additive models. J. Roy. Stat. Soc. Ser. B **71**(5), 1009–1030 (2009)
15. G. Raskutti, M.J. Wainwright, B. Yu, Restricted eigenvalue properties for correlated Gaussian designs. Annalen der Physik **322**(10), 891–921 (2010)
16. R. Tibshirani, M. Saunders, S. Rosset, J. Zhu, K. Knight, Sparsity and smoothness via the fused lasso. J Roy. Stat. Soc. Ser. B **67**(1), 91–108 (2005)
17. S.-A. van de Geer, High-dimensional generalized linear models and the lasso. Ann. Stat. **36**(2), 614–645 (2008)
18. Z. Wang, H. Liu, T. Zhang, Optimal computational and statistical rates of convergence for sparse nonconvex learning problems. Ann. Stat. **42**(6), 2164 (2014)

J. Lu, *Big Data Analysis*, https://doi.org/10.1007/978-3-032-03161-7

19. M. Yuan, Y. Lin, Model selection and estimation in regression with grouped variables. J. Roy. Stat. Soc. Ser. B **68**(1), 49–67 (2006)
20. H. Zou, The adaptive lasso and its oracle properties. J. Am. Stat. Assoc. **101**(476), 1418–1429 (2006)
21. H. Zou, T. Hastie, Regularization and variable selection via the elastic net. J. Roy. Stat. Soc. Ser. B **67**, 301–320 (2005)
22. R. Adamczak, A.E. Litvak, A. Pajor, N. Tomczak-Jaegermann, Quantitative estimations of the convergence of the empirical covariance matrix in log-concave ensembles. J. Am. Math. Soc. **23**, 535–561 (2010)
23. A. Agarwal, S. Negahban, M.J. Wainwright, Noisy matrix decomposition via convex relaxation: optimal rates in high dimensions. Ann. Stat. **40**(2), 1171–1197 (2012)
24. R. Ahlswede, A. Winter, Strong converse for identification via quantum channels. IEEE Trans. Inform. Theory **48**(3), 569–579 (2002)
25. M.A. Aizerman, E.M. Braverman, L.I. Rozonoer, Theoretical foundations of the potential function method in pattern recognition learning. Autom. Remote Control **25**, 821–837 (1964)
26. M. Akcakaya, V. Tarokh, Shannon theoretic limits on noisy compressive sampling. IEEE Trans. Inform. Theory **56**(1), 492–504 (2010)
27. K.S. Alexander, Rates of growth and sample moduli for weighted empirical processes indexed by sets. Probab. Theory Related Fields **75**, 379–423 (1987)
28. S. Alliney, S.A. Ruzinsky, An algorithm for the minimization of mixed ℓ_1 and ℓ_2 norms with application to Bayesian estimation. IEEE Trans. Signal Process. **42**(3), 618–627 (1994)
29. A.A. Amini, M.J. Wainwright, High-dimensional analysis of semdefinite relaxations for sparse principal component analysis. Ann. Stat. **5B**, 2877–2921 (2009)
30. A. Anandkumar, V.Y. F. Tan, F. Huang, A.S. Willsky, High-dimensional structure learning of Ising models: local separation criterion. Ann. Stat. **40**(3), 1346–1375 (2012)
31. T.W. Anderson, *An Introduction to Multivariate Statistical Analysis*. Wiley Series in Probability and Mathematical Statistics (Wiley, New York, 1984)
32. R.K. Ando, T. Zhang, A framework for learning predictive structures from multiple tasks and unlabeled data. J. Mach. Learn. Res. **6**, 1817–1853 (2005)
33. N. Aronszajn, Theory of reproducing kernels. Trans. Am. Math. Soc. **68**, 337–404 (1950)
34. P. Assouad, Deux remarques sur l'estimation. Comptes Rendus de l'Académie des Sciences, Paris **296**, 1021–1024 (1983)
35. K. Azuma, Weighted sums of certain dependent random variables. Tohoku Math. J. **19**, 357–367 (1967)
36. F. Bach, R. Jenatton, J. Mairal, G. Obozinski, Optimization with sparsity-inducing penalties. Found. Trends Mach. Learn. **4**(1), 1–106 (2012)
37. R.R. Bahadur, R.R. Rao, On deviations of the sample mean. Ann. Math. Stat. **31**, 1015–1027 (1960)
38. Z. Bai, J.W. Silverstein, *Spectral Analysis of Large Dimensional Random Matrices*. 2nd ed. (Springer, New York, 2010)
39. J. Baik, J.W. Silverstein, Eigenvalues of large sample covariance matrices of spiked populations models. J. Multivariate Anal. **97**(6), 1382–1408 (2006)
40. F. Balabdaoui, K. Rufibach, J.A. Wellner, Limit distribution theory for maximum likelihood estimation of a log-concave density. Ann. Stat. **62**(3), 1299–1331 (2009)
41. K. Ball, *An Elementary Introduction to Modern Convex Geometry*, in Flavors of Geometry. MSRI Publications, vol. 31 (Cambridge University Press, Cambridge, 1997), pp. 1–55
42. O. Banerjee, L. El Ghaoui, A. d'Aspremont, Model selection through sparse maximum likelihood estimation for multivariate Gaussian or binary data. J. Mach. Learn. Res. **9**, 485–516 (2008)
43. R.G. Baraniuk, V. Cevher, M.F. Duarte, C. Hegde, Model-based compressive sensing. IEEE Trans. Inform. Theory **56**(4), 1982–2001 (2010)
44. O.E. Barndorff-Nielson, *Information and Exponential Families* (Wiley, Chichester, 1978)
45. P.L. Bartlett, S. Mendelson, Gaussian and Rademacher complexities: risk bounds and structural results. J. Mach. Learn. Res. **3**, 463–482 (2002)

46. P.L. Bartlett, O. Bousquet, S. Mendelson, Local Rademacher complexities. Ann. Stat. **33**(4), 1497–1537 (2005)
47. R.J. Baxter, *Exactly Solved Models in Statistical Mechanics* (Academic Press, New York, 1982)
48. D. Bean, P.J. Bickel, N. El Karoui, B. Yu, Optimal M-estimation in high-dimensional regression. Proc. Natl. Acad. Sc. U.S.A. **110**(36), 14563–14568 (2013)
49. G. Bennett, Probability inequalities for the sum of independent random variables. J. Am. Stat. Assoc. **57**(297), 33–45 (1962)
50. J. Bento, A. Montanari, Which graphical models are difficult to learn? in *Proceedings of the NIPS Conference* (2009)
51. A. Berlinet, C. Thomas-Agnan, *Reproducing Kernel Hilbert Spaces in Probability and Statistics* (Kluwer Academic, Norwell, 2004)
52. S.N. Bernstein, On certain modifications of Chebyshev's inequality. Doklady Akademii Nauk SSSR **16**(6), 275–277 (1937)
53. Q. Berthet, P. Rigollet, Computational lower bounds for sparse PCA, in *Conference on Computational Learning Theory* (2013)
54. D.P. Bertsekas, *Convex Analysis and Optimization* (Athena Scientific, Boston, 2003)
55. J. Besag, Spatial interaction and the statistical analysis of lattice systems. J. Roy. Stat. Soc. Ser. B **36**, 192–236 (1974)
56. J. Besag, Statistical analysis of non-lattice data. Statistician **24**(3), 179–195 (1975)
57. J. Besag, Efficiency of pseudolikelihood estimation for simple Gaussian fields. Biometrika 64(3), 616–618 (1977)
58. H.A. Bethe, Statistics theory of superlattices. Proc. Roy. Soc. Lond. Ser. A **150**(871), 552–575 (1935)
59. R. Bhatia, *Matrix Analysis*. Graduate Texts in Mathematics (Springer, New York, 1997)
60. P.J. Bickel, K.A. Doksum, *Mathematical Statistics: Basic Ideas and Selected Topics* (CRC Press, Boca Raton, 2015)
61. P.J. Bickel, E. Levina, Covariance regularization by thresholding. Ann. Stat. **36**(6), 2577–2604 (2008)
62. P.J. Bickel, E. Levina, Regularized estimation of large covariance matrices. Ann. Stat. **36**(1), 199–227 (2008)
63. P.J. Bickel, Y. Ritov, A.B. Tsybakov, Simultaneous analysis of lasso and Dantzig selector. Ann. Stat. **37**(4), 1705–1732 (2009)
64. L. Birgé, Approximation dans les espaces metriques et theorie de l'estimation. Z. Wahrsch. verw. Gebiete **65**, 181–327 (1983)
65. L. Birgé, Estimating a density under order restrictions: non-asymptotic minimax risk. Ann. Stat. **15**(3), 995–1012 (1987)
66. L. Birgé, A new lower bound for multiple hypothesis testing. IEEE Trans. Inform. Theory **51**(4), 1611–1614 (2005)
67. L. Birgé, P. Massart, Estimation of integral functionals of a density. Ann. Stat. **23**(1), 11–29 (1995)
68. A. Birnbaum, I.M. Johnstone, B. Nadler, D. Paul, Minimax bounds for sparse PCA with noisy high-dimensional data. Ann. Stat. **41**(3), 1055–1084 (2012)
69. S.G. Bobkov, Isoperimetric and analytic inequalities for log-concave probability measures. Ann. Probab. **27**(4), 1903–1921 (1999)
70. S.G. Bobkov, F. Götze, Exponential integrability and transportation cost related to logarithmic Sobolev inequalities. J. Funct. Anal. **163**, 1–28 (1999)
71. S.G. Bobkov, M. Ledoux, From Brunn-Minkowski to Brascamp-Lieb and to logarithmic Sobolev inequalities. Geom. Funct. Anal. **10**, 1028–1052 (2000)
72. K. Borgwardt, A. Gretton, M. Rasch, H.P. Kriegel, B. Schölkopf, A.J. Smola, Integrating structured biological data by kernel maximum mean discrepancy. Bioinformatics **22**(14), 49–57 (2006)
73. J. Borwein, A. Lewis, *Convex Analysis* (Springer, New York, 1999)

74. B.E. Boser, I.M. Guyon, V.N. Vapnik, A training algorithm for optimal margin classifiers, in *Proceedings of the Conference on Learning Theory (COLT)* (ACM, New York, 1992), pp. 144–152
75. S. Boucheron, G. Lugosi, P. Massart, Concentration inequalities using the entropy method. Ann. Probab. **31**(3), 1583–1614 (2003)
76. S. Boucheron, G. Lugosi, P. Massart, *Concentration Inequalities: A Nonasymptotic Theory of Independence* (Oxford University Press, Oxford, 2013)
77. J. Bourgain, S. Dirksen, J. Nelson, Toward a unified theory of sparse dimensionality reduction in Euclidean space. Geom. Funct. Anal. **25**(4) (2015)
78. O. Bousquet, A Bennett concentration inequality and its application to suprema of empirical processes. Comptes Rendus de l'Académie des Sciences, Paris, Série I **334**, 495–500 (2002)
79. O. Bousquet, Concentration inequalities for sub-additive functions using the entropy method. Stoch. Inequalities Appl. **56**, 213–247 (2003)
80. S. Boyd, L. Vandenberghe, *Convex Optimization* (Cambridge University Press, Cambridge, 2004)
81. H.J. Brascamp, E.H. Lieb, On extensions of the Brunn–Minkowski and Prékopa–Leindler theorems, including inequalities for log concave functions, and with an application to the diffusion equation. J. Funct. Anal. **22**, 366–389 (1976)
82. L. Breiman, *Probability*. Classics in Applied Mathematics (SIAM, Philadelphia, 1992)
83. G. Bresler, Efficiently learning Ising models on arbitrary graphs. Tech. rept. MIT (2014)
84. G. Bresler, E. Mossel, A. Sly, Reconstruction of Markov random fields from samples: some observations and algorithms. SIAM J. Comput. **42**(2), 563–578 (2013)
85. E.M. Bronshtein: ϵ-entropy of convex sets and functions. Siberian Math. J. **17**, 393–398 (1976)
86. L.D. Brown, *Fundamentals of statistical exponential families* (Institute of Mathematical Statistics, Hayward, 1986)
87. H.D. Brunk, Maximum likelihood estimates of monotone parameters. Ann. Math. Stat. **26**, 607–616 (1955)
88. H.D. Brunk, *Estimation of Isotonic Regression*, in Nonparametric Techniques in Statistical Inference (Cambridge University Press, New York, 1970), pp. 177–197
89. P. Bühlmann, S. van de Geer, *Statistics for High-Dimensional Data*. Springer Series in Statistics (Springer, Berlin, 2011)
90. A. Buja, T.J. Hastie, R. Tibshirani, Linear smoothers and additive models. Ann. Stat. **17**(2), 453–510 (1989)
91. V.V. Buldygin, Y.V. Kozachenko, *Metric Characterization of Random Variables and Random Processes* (American Mathematical Society, Providence, 2000)
92. F. Bunea, A.B. Tsybakov, M. Wegkamp, Sparsity oracle inequalities for the Lasso. Electron. J. Stat. **1**, 169–194 (2007)
93. F. Bunea, Y. She, M. Wegkamp, Optimal selection of reduced rank estimators of high-dimensional matrices. Ann. Stat. **39**(2), 1282–1309 (2011)
94. T.T. Cai, C.H. Zhang, H.H. Zhou, Optimal rates of convergence for covariance matrix estimation. Ann. Stat. **38**(4), 2118–2144 (2010)
95. T.T. Cai, W. Liu, X. Luo, A constrained ℓ_1-minimization approach to sparse precision matrix estimation. J. Am. Stat. Assoc. **106**, 594–607 (2011)
96. T.T. Cai, T. Liang, A. Rakhlin, Computational and statistical boundaries for submatrix localization in a large noisy matrix. Tech. rept. Univ. Penn. (2015)
97. E.J. Candès, Y. Plan, Matrix completion with noise. Proc. IEEE **98**(6), 925–936 (2010)
98. E.J. Candès, B. Recht, Exact matrix completion via convex optimization. Found. Comput. Math. **9**(6), 717–772 (2009)
99. E.J. Candès, T. Tao, Decoding by linear programming. IEEE Trans. Inform. Theory **51**(12), 4203–4215 (2005)
100. E.J. Candès, T. Tao, The Dantzig selector: statistical estimation when p is much larger than n. Ann. Stat. **35**(6), 2313–2351 (2007)

101. E.J. Candès, X. Li, Y. Ma, J. Wright, Robust principal component analysis? J. ACM **58**(3), 11 (37pp) (2011)
102. E.J. Candès, T. Strohmer, V. Voroninski, PhaseLift: exact and stable signal recovery from magnitude measurements via convex programming. Commun. Pure Appl. Math. **66**(8), 1241–1274 (2013)
103. F.P. Cantelli, Sulla determinazione empirica della legge di probabilita. Giornale dell'Istituto Italiano degli Attuari **4**, 421–424 (1933)
104. B. Carl, A. Pajor, Gelfand numbers of operators with values in a Hilbert space. Inventiones Mathematicae **94**, 479–504 (1988)
105. B. Carl, I. Stephani, *Entropy, Compactness and the Approximation of Operators*. Cambridge Tracts in Mathematics (Cambridge University Press, Cambridge, 1990)
106. E. Carlen, Trace inequalities and quantum entropy: an introductory course, in *Entropy and the Quantum* (American Mathematical Society, Providence, 2009)
107. R.J. Carroll, D. Ruppert, L.A. Stefanski, *Measurement Error in Nonlinear Models* (Chapman & Hall/CRC, Boca Raton, 1995)
108. A. Chai, M. Moscoso, G. Papanicolaou, Array imaging using intensity-only measurements. Inverse Problems **27**(1), 1–15 (2011)
109. V. Chandrasekaran, S. Sanghavi, P.A. Parrilo, A.S. Willsky, Rank-sparsity incoherence for matrix decomposition. SIAM J. Optim. **21**, 572–596 (2011)
110. V. Chandrasekaran, B. Recht, P.A. Parrilo, A.S. Willsky, The convex geometry of linear inverse problems. Found. Comput. Math. **12**(6), 805–849 (2012)
111. V. Chandrasekaran, P.A. Parrilo, A.S. Willsky, Latent variable graphical model selection via convex optimization. Ann. Stat. **40**(4), 1935–1967 (2012)
112. S. Chatterjee, An error bound in the Sudakov-Fernique inequality. Tech. rept. UC Berkeley. arXiv:math.PR/0510424 (2005)
113. S. Chatterjee, Stein's method for concentration inequalities. Probab. Theory Related Fields **138**(1–2), 305–321 (2007)
114. S. Chatterjee, A. Guntuboyina, B. Sen, On risk bounds in isotonic and other shape restricted regression problems. Ann. Stat. **43**(4), 1774–1800 (2015)
115. S. Chen, D.L. Donoho, M.A. Saunders, Atomic decomposition by basis pursuit. SIAM J. Sci. Comput. **20**(1), 33–61 (1998)
116. H. Chernoff, A measure of asymptotic efficiency for tests of a hypothesis based on a sum of observations. Ann. Math. Stat. **23**, 493–507 (1952)
117. V. Chernozhukov, D. Chetverikov, K. Kato, Comparison and anti-concentration bounds for maxima of Gaussian random vectors. Tech. rept. MIT (2013)
118. F.R.K. Chung, *Spectral Graph Theory* (American Mathematical Society, Providence, 1991)
119. P. Clifford, Markov random fields in statistics, in *Disorder in Physical Systems*, ed. by G.R. Grimmett, D.J.A. Welsh (Oxford Science Publications, 1990)
120. A. Cohen, W. Dahmen, R.A. DeVore, Compressed sensing and best k-term approximation. J. Am. Math. Soc. **22**(1), 211–231 (2008)
121. G. Cormode, Synopses for massive data: samples, histograms, wavelets and sketches. Found. Trends Databases **4**(2), 1–294 (2012)
122. T.M. Cover, J.A. Thomas, *Elements of Information Theory* (Wiley, New York, 1991)
123. M. Cule, R.J. Samworth, M. Stewart, Maximum likelihood estimation of a multi-dimensional log-concave density. J. R. Stat. Soc. B **62**, 545–607 (2010)
124. A.S. Dalalyan, M. Hebiri, J. Lederer, On the prediction performance of the Lasso. Tech. rept. ENSAE. arxiv:1402,1700, to appear in Bernoulli (2014)
125. A. d'Aspremont, L. El Ghaoui, M.I. Jordan, G.R. Lanckriet, A direct formulation for sparse PCA using semidefinite programming. SIAM Rev. **49**(3), 434–448 (2007)
126. A. d'Aspremont, O. Banerjee, L. El Ghaoui, First order methods for sparse covariance selection. SIAM J. Matrix Anal. Appl. **30**(1), 55–66 (2008)
127. K.R. Davidson, S.J. Szarek, *Local Operator Theory, Random Matrices, and Banach Spaces*, in *Handbook of Banach Spaces*, vol. 1 (Elsevier, Amsterdam, 2001), pp. 317–336
128. A.P. Dawid, The geometry of proper scoring rules. Ann. Inst. Stat. Math. **59**, 77–93 (2007)

129. V. de La Pena, E. Giné, *Decoupling: From Dependence to Independence* (Springer, New York, 1999)
130. A. Dembo, Information inequalities and concentration of measure. Ann. Probab. **25**(2), 927–939 (1997)
131. A. Dembo, O. Zeitouni, Transportation approach to some concentration inequalities in product spaces. Electron. Commun. Probab. **1**, 83–90 (1996)
132. R.A. DeVore, G.G. Lorentz, *Constructive Approximation* (Springer, New York, 1993)
133. D.L. Donoho, For most large underdetermined systems of linear equations, the minimal ℓ_1-norm near-solution approximates the sparsest near-solution. Commun. Pure Appl. Math. **59**(7), 907–934 (2006)
134. D.L. Donoho, For most large underdetermined systems of linear equations, the minimal ℓ_1-norm solution is also the sparsest solution. Commun. Pure Appl. Math. **59**(6), 797–829 (2006)
135. D.L. Donoho, X. Huo, Uncertainty principles and ideal atomic decomposition. IEEE Trans. Inform. Theory **47**(7), 2845–2862 (2001)
136. D.L. Donoho, I.M. Johnstone, Minimax risk over ℓ_p-balls for ℓ_q-error. Probab. Theory Related Fields **99**, 277–303 (1994)
137. D.L. Donoho, A. Montanari, High dimensional robust M-estimation: asymptotic variance via approximate message passing. Tech. rept. Stanford University. Posted as arxiv:1310.7320 (2013)
138. D.L. Donoho, P.B. Stark, Uncertainty principles and signal recovery. SIAM J. Appl. Math. **49**, 906–931 (1989)
139. D.L. Donoho, J.M. Tanner, Counting faces of randomly-projected polytopes when the projection radically lowers dimension. J. Am. Math. Soc. **22**(1), 1–53 (2009)
140. J.C. Duchi, M.J. Wainwright, M.I. Jordan, Local privacy and minimax bounds: sharp rates for probability estimation. Tech. rept. UC Berkeley (2013)
141. J.C. Duchi, M.J. Wainwright, M.I. Jordan, Privacy-aware learning. J. ACM, 61(6), 37 (2014)
142. R.M. Dudley, The sizes of compact subsets of Hilbert spaces and continuity of Gaussian processes. J. Funct. Anal. **1**, 290–330 (1967)
143. R.M. Dudley, Central limit theorems for empirical measures. Ann. Probab. **6**, 899–929 (1978)
144. R.M. Dudley, *Uniform Central Limit Theorems* (Cambridge University Press, Cambridge, 1999)
145. L.Dümbgen, R.J. Samworth, D. Schuhmacher, Approximation by log-concave distributions with applications to regression. Ann. Stat. **39**(2), 702–730 (2011)
146. R. Durrett, *Probability: Theory and Examples* (Cambridge University Press, Cambridge, 2010)
147. A. Dvoretsky, J. Kiefer, J. Wolfowitz, Asymptotic minimax character of the sample distribution function and of the classical multinomial estimator. Ann. Math. Stat. **27**, 642–669 (1956)
148. P.P.B. Eggermont, V.N. LaRiccia, *Maximum Penalized Likelihood Estimation: V. I Density Estimation*. Springer Series in Statistics, vol. 1 (Springer, New York, 2001)
149. P.P.B. Eggermont, V.N. LaRiccia, *Maximum Penalized Likelihood Estimation: V. II Regression*. Springer Series in Statistics, vol. 2 (Springer, New York, 2007)
150. N. El Karoui, Operator norm consistent estimation of large-dimensional sparse covariance matrices. Ann. Stat. **36**(6), 2717–2756 (2008)
151. N. El Karoui, Asymptotic behavior of unregularized and ridge-regularized high-dimensional robust regression estimators: rigorous results. Tech. rept. UC Berkeley. Posted as arxiv:1311.2445 (2013)
152. N. El Karoui, D. Bean, P.J. Bickel, B. Yu, On robust regression with high-dimensional predictors. Proc. Natl. Acad. Sci. U.S.A. **110**(36), 14557–14562 (2013)
153. M. Elad, A.M. Bruckstein, A generalized uncertainty principle and sparse representation in pairs of bases. IEEE Trans. Inform. Theory **48**(9), 2558–2567 (2002)
154. J. Fan, J. Lv, Nonconcave penalized likelihood with NP-dimensionality. IEEE Trans. Inform. Theory **57**(8), 5467–5484 (2011)
155. J. Fan, Y. Liao, M. Mincheva, Large covariance estimation by thresholding principal orthogonal components. J. Roy. Stat. Soc. B **75**, 603–680 (2013)

156. J. Fan, L. Xue, H. Zou, Strong oracle optimality of folded concave penalized estimation. Ann. Stat. **42**(3), 819–849 (2014)
157. M. Fazel, Matrix Rank Minimization with Applications. Ph.D. Thesis, Stanford, 2002. http://faculty.washington.edu/mfazel/thesis-final.pdf
158. X.M. Fernique, Des resultats nouveaux sur les processus Gaussiens. Comptes Rendus de l'Académie des Sciences, Paris **278**, A363–A365 (1974)
159. A. Feuer, A. Nemirovski, On sparse representation in pairs of bases. IEEE Trans. Inform. Theory **49**(6), 1579–1581 (2003)
160. J.R. Fienup, Phase retrieval algorithms: a comparison. Appl. Opt. **21**(15), 2758–2769 (1982)
161. J.R. Fienup, C.C. Wackerman, Phase-retrieval stagnation problems and solutions. J. Opt. Soc. Am. A **3**, 1897–1907 (1986)
162. A.K. Fletcher, S. Rangan, V.K. Goyal, Necessary and sufficient conditions for sparsity pattern recovery. IEEE Trans. Inform. Theory **55**(12), 5758–5772 (2009)
163. R. Foygel, N. Srebro, Fast rate and optimistic rate for ℓ_1-regularized regression. Tech. rept. Toyoto Technological Institute. arXiv:1108.037v1 (2011)
164. J.H. Friedman, W. Stuetzle, Projection pursuit regression. J. Am. Stat. Assoc. **76**(376), 817–823 (1981)
165. J.H. Friedman, J.W. Tukey, A projection pursuit algorithm for exploratory data analysis. IEEE Trans. Comput. **C-23**, 881–889 (1994)
166. J.H. Friedman, T.J. Hastie, R. Tibshirani, Sparse inverse covariance estimation with the graphical Lasso. Biostatistics **9**(3), 432–441 (2008)
167. J.J. Fuchs, *Recovery of Exact Sparse Representations in the Presence of Noise*, in *ICASSP*, vol. 2 (2004)
168. R.G. Gallager, *Information Theory and Reliable Communication* (Wiley, New York, 1968)
169. C. Gao, Z. Ma, H.H. Zhou, Sparse CCA: adaptive estimation and computational barriers. Tech. rept. Yale University (2015)
170. R.J. Gardner, The Brunn-Minkowski inequality. Bull. Am. Math. Soc. **39**, 355–405 (2002)
171. S. Geman, A limit theorem for the norm of random matrices. Ann. Probab. **8**(2), 252–261 (1980)
172. S. Geman, D. Geman, Stochastic relaxation, Gibbs distributions, and the Bayesian restoration of images. IEEE Trans. Pattern Anal. Mach. Intell. **6**, 721–741 (1984)
173. S. Geman, C.R. Hwang, Nonparametric maximum likelihood estimation by the method of sieves. Ann. Stat. **10**(2), 401–414 (1982)
174. V. Glivenko, Sulla determinazione empirica della legge di probabilita. Giornale dell'Istituto Italiano degli Attuari **4**, 92–99 (1933)
175. T. Gneiting, A.E. Raftery, Strictly proper scoring rules, prediction, and estimation. J. Am. Stat. Assoc. **102**(477), 359–378 (2007)
176. K. Goldberg, T. Roeder, D. Gupta, C. Perkins, Eigentaste: a constant time collaborative filtering algorithm. Inform. Retrieval **4**(2), 133–151 (2001)
177. I.J. Good, R.A. Gaskins, Nonparametric roughness penalties for probability densities. Biometrika **58**, 255–277 (1971)
178. Y. Gordon, Some inequalities for Gaussian processes and applications. Israel J. Math. **50**, 265–289 (1985)
179. Y. Gordon, On Milman's inequality and random subspaces which escape through a mesh in $\mathbb{R}^n$, in *Geometric Aspects of Functional Analysis*. Lecture Notes in Mathematics, vol. 1317 (Springer, Berlin, 1986), pp. 84–106
180. Y. Gordon, Elliptically contoured distributions. Probab. Theory Related Fields **76**, 429–438 (1987)
181. F. Götze, A. Tikhomirov, Rate of convergence in probability to the Marčenko-Pastur law. Bernoulli **10**(3), 503–548 (2004)
182. R.W. Grechberg, W.O. Saxton, A practical algorithm for the determination of phase from image and diffraction plane intensities. Optik **35**, 237–246 (1972)
183. E. Greenshtein, Y. Ritov, Persistency in high dimensional linear predictor-selection and the virtue of over-parametrization. Bernoulli **10**, 971–988 (2004)

184. A. Gretton, K. Borgwardt, M. Rasch, B. Schölkopf, A. Smola, A kernel two-sample test. J. Mach. Learn. Res. **13**, 723–773 (2012)
185. D. Griffin, J. Lim, Signal estimation from modified short-time Fourier transforms. IEEE Trans. Acoust. Speech Signal Process. **32**(2), 236–243 (1984)
186. G.R. Grimmett, A theorem about random fields. Bull. Lond. Math. Soc. **5**, 81–84 (1973)
187. D. Gross, Recovering low-rank matrices from few coefficients in any basis. IEEE Trans. Inform. Theory **57**(3), 1548–1566 (2011)
188. L. Gross, Logarithmic Sobolev inequalities. Am. J. Math. **97**, 1061–1083 (1975)
189. C. Gu, *Smoothing Spline ANOVA Models*. Springer Series in Statistics (Springer, New York, 2002)
190. O. Guédon, A.E. Litvak, *Euclidean Projections of a p-Convex Body*, in Geometric Aspects of Functional Analysis (Springer, Berlin, 2000), pp. 95–108
191. A. Guntuboyina, Lower bounds for the minimax risk using f-divergences and applications. IEEE Trans. Inform. Theory **57**(4), 2386–2399 (2011)
192. A. Guntuboyina, B. Sen, Covering numbers for convex functions. IEEE Trans. Inform. Theory **59**, 1957–1965 (2013)
193. L. Gyorfi, M. Kohler, A. Krzyzak, H. Walk, *A Distribution-Free Theory of Nonparametric Regression*. Springer Series in Statistics (Springer, Berlin, 2002)
194. J.M. Hammersley, P. Clifford, Markov fields on finite graphs and lattices. Unpublished (1971)
195. D.L. Hanson, G. Pledger, Consistency in concave regression. Ann. Stat. **4**, 1038–1050 (1976)
196. D.L. Hanson, F.T. Wright, A bound on tail probabilities for quadratic forms in independent random variables. Ann. Math. Stat. **42**(3), 1079–1083 (1971)
197. W.K. Härdle, T.M. Stoker, Investigating smooth multiple regression by the method of average derivatives. J. Am. Stat. Assoc. **84**, 986–995 (1989)
198. W.K. Härdle, P. Hall, H. Ichimura, Optimal smoothing in single-index models. Ann. Stat. **21**, 157–178 (1993)
199. W.K. Härdle, M. Müller, S. Sperlich, A. Werwatz, *Nonparametric and Semiparametric Models*. Springer Series in Statistics (Springer, New York, 2004)
200. L.H. Harper, Optimal numberings and isoperimetric problems on graphs. J. Comb. Theory **1**, 385–393 (1966)
201. R.W. Harrison, Phase problem in crystallography. J. Opt. Soc. Am. A **10**(5), 1046–1055 (1993)
202. R.Z. Hasminskii, A lower bound on the risks of nonparametric estimates of densities in the uniform metric. Theory Probab. Appl. **23**, 794–798 (1978)
203. R.Z. Hasminskii, I. Ibragimov, *Statistical Estimation: Asymptotic Theory* (Springer, New York, 1981)
204. R.Z. Hasminskii, I. Ibragimov, On density estimation in the view of Kolmogorov's ideas in approximation theory. Ann. Stat. **18**(3), 999–1010 (1990)
205. T.J. Hastie, R. Tibshirani, Generalized additive models. Stat. Sci. **1**(3), 297–310 (1986)
206. T.J. Hastie, R. Tibshirani, *Generalized Additive Models* (Chapman & Hall/CRC, Boca Raton, 1990)
207. C. Hildreth, Point estimates of ordinates of concave functions. J. Am. Stat. Assoc. **49**, 598–619 (1954)
208. J. Hiriart-Urruty, C. Lemaréchal, *Convex Analysis and Minimization Algorithms*, vol. 1 (Springer, New York, 1993)
209. W. Hoeffding, Probability inequalities for sums of bounded random variables. J. Am. Stat. Assoc. **58**, 13–30 (1963)
210. A.E. Hoerl, R.W. Kennard, Ridge regression: biased estimation for nonorthogonal problems. Technometrics **12**, 55–67 (1970)
211. H. Hölfing, R. Tibshirani, Estimation of sparse binary pairwise Markov networks using pseudo-likelihoods. J. Mach. Learn. Res. **19**, 883–906 (2009)
212. R. Holley, D. Stroock, Log Sobolev inequalities and stochastic Ising models. J. Stat. Phys. **46**(5), 1159–1194 (1987)
213. R.A. Horn, C.R. Johnson, *Matrix Analysis* (Cambridge University Press, Cambridge, 1985)

214. R.A. Horn, C.R. Johnson, *Topics in Matrix Analysis* (Cambridge University Press, Cambridge, 1991)
215. M. Hristache, A. Juditsky, V. Spokoiny, Direct estimation of the index coefficient in a single index model. Ann. Stat. **29**, 595–623 (2001)
216. D. Hsu, S.M. Kakade, T. Zhang, Tail inequalities for sums of random matrices that depend on the intrinsic dimension. Electron. Commun. Probab. **17**(14), 1–13 (2012)
217. D. Hsu, S.M. Kakade, T. Zhang, A tail inequality for quadratic forms of sub-Gaussian random vectors. Electron. J. Probab. **52**, 1–6 (2012)
218. J. Huang, T. Zhang, The benefit of group sparsity. Ann. Stat. **38**(4), 1978–2004 (2010)
219. J. Huang, S. Ma, C.H. Zhang, Adaptive Lasso for sparse high-dimensional regression models. Statistica Sinica **18**, 1603–1618 (2008)
220. P.J. Huber, Robust regression: asymptotics, conjectures and Monte Carlo. Ann. Stat. **1**(5), 799–821 (1973)
221. P.J. Huber, Projection pursuit. Ann. Stat. **13**(2), 435–475 (1985)
222. H. Ichimura, Semiparametric least squares (SLS) and weighted (SLS) estimation of single index models. J. Econom. **58**, 71–120 (1993)
223. E. Ising, Beitrag zur Theorie der Ferromagnetismus. Zeitschrift für Physik **31**(1), 253–258 (1925)
224. S.J. Iturria, R.J. Carroll, D. Firth, Polynomial regression and estimating functions in the presence of multiplicative measurement error. J. Roy. Stat. Soc. B **61**, 547–561 (1999)
225. A.J. Izenman, Reduced-rank regression for the multivariate linear model. J. Multivariate Anal. **5**, 248–264 (1975)
226. A.J. Izenman, *Modern Multivariate Statistical Techniques: Regression, Classification and Manifold Learning* (Springer, New York, 2008)
227. L. Jacob, G. Obozinski, J.P. Vert, Group Lasso with overlap and graph Lasso, in, *International Conference on Machine Learning (ICML)* (2009), pp. 433–440
228. A. Jalali, P. Ravikumar, S. Sanghavi, C. Ruan, A dirty model for multi-task learning, in *Advances in Neural Information Processing Systems*, vol. 23 (2010), pp. 964–972
229. W.B. Johnson, J. Lindenstrauss, Extensions of Lipschitz mappings into a Hilbert space. Contemp. Math. **26**, 189–206 (1984)
230. I.M. Johnstone, On the distribution of the largest eigenvalue in principal components analysis. Ann. Stat. **29**(2), 295–327 (2001)
231. I.M. Johnstone, *Gaussian Estimation: Sequence and Wavelet Models* (Springer, New York, 2015)
232. I.M. Johnstone, A. Y. Lu, On consistency and sparsity for principal components analysis in high dimensions. J. Am. Stat. Assoc. **104**, 682–693 (2009)
233. I.T. Jolliffe, *Principal Component Analysis* (Springer, New York, 2004)
234. I.T. Jolliffe, N.T. Trendafilov, M. Uddin, A modified principal component technique based on the LASSO. J. Comput. Graph. Stat. **12**, 531–547 (2003)
235. A. Juditsky, A. Nemirovski, Functional aggregation for nonparametric regression. Ann. Stat. **28**, 681–712 (2000)
236. J.P. Kahane, Une inequalité du type de Slepian et Gordon sur les processus Gaussiens. Israel J. Math. **55**, 109–110 (1986)
237. M. Kalisch, P. Bühlmann, Estimating high-dimensional directed acyclic graphs with the PC algorithm. J. Mach. Learn. Res. **8**, 613–636 (2007)
238. D.M. Kane, J. Nelson, Sparser Johnson-Lindenstrauss transforms. J. ACM **61**(1), 1–23 (2014)
239. L.V. Kantorovich, G.S. Rubinstein, On the space of completely additive functions. Vestnik Leningrad Univ. Ser. Math. Mekh. i. Astron **13**(7), 52–59 (1958)
240. R.W. Keener, *Theoretical Statistics: Topics for a Core Class* (Springer, New York, 2010)
241. R.H. Keshavan, A. Montanari, S. Oh, Matrix completion from few entries. IEEE Trans. Inform. Theory **56**(6), 2980–2998 (2010)
242. R.H. Keshavan, A. Montanari, S. Oh, Matrix completion from noisy entries. J. Mach. Learn. Res. **11**, 2057–2078 (2010)
243. Y. Kim, J. Kim, Y. Kim, Blockwise sparse regression. Statistica Sinica **16**(2), 375–390 (2006)

244. G. Kimeldorf, G. Wahba, Some results on Tchebycheffian spline functions. J. Math. Anal. Appl. **33**, 82–95 (1971)
245. T. Klein, E. Rio, Concentration around the mean for maxima of empirical processes. Ann. Probab. **33**(3), 1060–1077 (2005)
246. D. Koller, N. Friedman, *Graphical Models* (MIT Press, New York, 2010)
247. A.N. Kolmogorov, Asymptotic characterization of some completely bounded metric spaces. Doklady Akademii Nauk SSSR **108**, 585–589 (1956)
248. A.N. Kolmogorov, Linear dimension of topological vector spaces. Doklady Akademii Nauk SSSR **120**, 239–241 (1958)
249. A.N. Kolmogorov, B. Tikhomirov, ϵ-entropy and ϵ-capacity of sets in functional spaces. Uspekhi Mat. Nauk. **86**, 3–86 (1959)
250. V. Koltchinskii, Rademacher penalities and structural risk minimization. IEEE Trans. Inform. Theory **47**(5), 1902–1914 (2001)
251. V. Koltchinskii, Local Rademacher complexities and oracle inequalities in risk minimization. Ann. Stat. **34**(6), 2593–2656 (2006)
252. V. Koltchinskii, D. Panchenko, Rademacher processes and bounding the risk of function learning, in *High-Dimensional Probability II* (Springer, Berlin, 2000), pp. 443–459
253. V. Koltchinskii, M. Yuan, Sparsity in multiple kernel learning. Ann. Stat. **38**, 3660–3695 (2010)
254. V. Koltchinskii, K. Lounici, A.B. Tsybakov, Nuclear-norm penalization and optimal rates for noisy low-rank matrix completion. Ann. Stat. **39**, 2302–2329 (2011)
255. L.A. Kontorovich, K. Ramanan, Concentration inequalities for dependent random variables via the martingale method. Ann. Probab. **36**(6), 2126–2158 (2008)
256. J.B. Kruskal, Towards a practical method which helps uncover the structure of a set of multivariate observation by finding the linear transformation which optimizes a new 'index of condensation', in *Statistical Computation* (Academic Press, New York, 1969)
257. T. Kühn, A lower estimate for entropy numbers. J. Approx. Theory **110**, 120–124 (2001)
258. S. Kullback, R.A. Leibler, On information and sufficiency. Ann. Math. Stat. **22**(1), 79–86 (1951)
259. C. Lam, J. Fan, Sparsistency and rates of convergence in large covariance matrix estimation. Ann. Stat. **37**, 4254–4278 (2009)
260. M. Laurent, Matrix completion problems, in *The Encyclopedia of Optimization* (Kluwer Academic, Dordrecht, 2001), pp. 221–229
261. M. Laurent, A comparison of the Sherali-Adams, Lovász-Schrijver and Lasserre relaxations for 0–1 programming. Math. Oper. Res. **28**, 470–496 (2003)
262. S.L. Lauritzen, *Graphical Models* (Oxford University Press, Oxford, 1996)
263. L. Le Cam, Convergence of estimates under dimensionality restrictions. Ann. Stat. . **1**(1), 38–53 (1973)
264. M. Ledoux, On Talagrand's deviation inequalities for product measures. ESAIM: Probab. Stat. **1**, 63–87 (1996)
265. M. Ledoux, *The Concentration of Measure Phenomenon*. Mathematical Surveys and Monographs (American Mathematical Society, Providence, 2001)
266. M. Ledoux, M. Talagrand, *Probability in Banach Spaces: Isoperimetry and Processes* (Springer, New York, 1991)
267. J.D. Lee, Y. Sun, J. Taylor, On model selection consistency of M-estimators with geometrically decomposable penalties. Tech. rept. Stanford University. arxiv1305.7477v4 (2013)
268. L. Leindler, On a certain converse of Hölder's inequality. Acta Scientiarum Mathematicarum (Szeged) **33**, 217–223 (1972)
269. S. Levy, P.K. Fullagar, Reconstruction of a sparse spike train from a portion of its spectrum and application to high-resolution deconvolution. Geophysics **46**(9), 1235–1243 (1981)
270. E.H. Lieb, Convex trace functions and the Wigner-Yanase-Dyson conjecture. Adv. Math. **11**, 267–288 (1973)
271. D.V. Lindley, On a measure of the information provided by an experiment. Ann. Math. Stat. **27**, 902–908 (1956)

272. H. Liu, J.D. Lafferty, L.A. Wasserman, The nonparanormal: Semiparametric estimation of high-dimensional undirected graphs. J. Mach. Learn. Res. **10**, 1–37 (2009)
273. H. Liu, F. Han, M. Yuan, J.D. Lafferty, L.A. Wasserman, High-dimensional semiparametric Gaussian copula graphical models. Ann. Stat. **40**(4), 2293–2326 (2012)
274. P. Loh, M.J. Wainwright, High-dimensional regression with noisy and missing data: provable guarantees with non-convexity. Ann. Stat. **40**(3), 1637–1664 (2012)
275. P. Loh, M.J. Wainwright, Structure estimation for discrete graphical models: generalized covariance matrices and their inverses. Ann. Stat. **41**(6), 3022–3049 (2013)
276. P. Loh, M.J. Wainwright, Regularized M-estimators with nonconvexity: statistical and algorithmic theory for local optima. J. Mach. Learn. Res. **16**, 559–616 (2015)
277. P. Loh, M.J. Wainwright, Support recovery without incoherence: a case for nonconvex regularization. Ann. Stat. **45**(6), 2455–2482 (2017)
278. K. Lounici, High-dimensional covariance matrix estimation with missing observations. Bernoulli **20**(3), 1029–1058 (2014)
279. K. Lounici, M. Pontil, S. van de Geer, A.B. Tsybakov, Oracle inequalities and optimal inference under group sparsity. Ann. Stat. **39**(4), 2164–2204 (2011)
280. L. Lovász, M.D. Plummer, *Matching Theory*. Annals of Discrete Mathematics (North-Holland Mathematics Studies, 1986)
281. D.G. Luenberger, *Optimization by Vector Space Methods* (Wiley, New York, 1969)
282. Z.Q. Luo, Universal decentralized estimation in a bandwidth constrained sensor network. IEEE Trans. Inform. Theory **51**(6), 2210–2219 (2005)
283. Z.Q. Luo, J.F. Sturm, S. Zhang, Multivariate nonnegative quadratic mappings. SIAM J. Optim. **14**(4), 1140–1162 (2004)
284. K.V. Mardia, J.T. Kent, J.M. Bibby, *Multivariate Analysis* (Academic Press, London, 1979)
285. V.A. Marčenko, L.A. Pastur, Distribution of eigenvalues for some sets of random matrices. Math. USSR-Sbornik **1**(4), 457–483 (1967)
286. R. Martin, R. Mess, Concentration inequalities for rectangular matrix products. Tech. rept. Statistics Department, Stanford University (2015)
287. P. Massart, About the constants in Talagrand's concentration inequalities for empirical processes. Ann. Probab. **28**(2), 863–884 (2000)
288. P. Massart, Some applications of concentration inequalities to statistics. Annales de la Faculté des Sciences de Toulouse **IX**, 245–303 (2000)
289. P. Massart, *Concentration Inequalities and Model Selection*. Ecole d'Eté de Probabilités de Saint-Flour XXXIII (Springer, New York, 2003)
290. P. Massart, E. Nédélec, Risk bounds for statistical learning. Ann. Stat. **34**(5), 2326–2366 (2006)
291. C. McDiarmid, On the method of bounded differences, in *Surveys in Combinatorics*. London Mathematical Society Lecture Note Series, vol. 141 (Cambridge University Press, Cambridge, 1989), pp. 148–188
292. N. Meinshausen, P. Bühlmann, High dimensional graphs and variable selection with the Lasso. Ann. Stat. **34**, 1436–1462 (2006)
293. N. Meinshausen, B. Yu, Lasso-type recovery of sparse representations for high-dimensional data. Ann. Stat. **37**(1), 246–270 (2009)
294. S. Mendelson, Geometric parameters of kernel machines, in *Proceedings of COLT* (2002), pp. 29–43
295. S. Mendelson, A few notes on statistical learning theory, in *Advanced Lectures in Machine Learning*. Lecture Notes in Computer Science, vol. 2600 (Springer, New York, 2003), pp. 1–40
296. S. Mendelson, Lower bounds for the empirical minimization algorithm. IEEE Trans. Inform. Theory **54**(8), 3797–3803 (2008)
297. S. Mendelson, Upper bounds on product and multiplier empirical processes. Stoch. Process. Appl. **126**(12), 3652–3680 (2016)
298. R. von Mises, *Mathematical Theory of Probability and Statistics* (Academic Press, New York, 1964)

299. A. Montanari, Graphical models concepts in compressed sensing, in *Compressed Sensing: Theory and Applications* (Cambridge University Press, Cambridge, 2012)
300. A. Montanari, E. Richard, Non-negative principal component analysis: message passing algorithms and sharp asymptotics. IEEE Trans. Inform. Theory **62**(3), 1458–1484 (2016)
301. J.J. Moreau, Proximité et dualité dans un espace Hilbertien. Bull. Soc. Math. France **93**, 273–299 (1965)
302. E. Mossel, S. Roch, Learning nonsingular phylogenies and hidden Markov models. Ann. Appl. Probab. **16**(2), 583–614 (2006)
303. R.J. Muirhead, *Aspects of Multivariate Statistical Theory* (Wiley, New York, 1982)
304. B.K. Natarajan, Sparse approximate solutions to linear systems. SIAM J. Comput. **24**(2), 227–234 (1995)
305. S. Negahban, M.J. Wainwright, Estimation of (near) low-rank matrices with noise and high-dimensional scaling. Ann. Stat. **39**(2), 1069–1097 (2011)
306. S. Negahban, M.J. Wainwright, Restricted strong convexity and (weighted) matrix completion: optimal bounds with noise. J. Mach. Learn. Res. **13**, 1665–1697 (2012)
307. S. Negahban, P. Ravikumar, M.J. Wainwright, B. Yu, A unified framework for high-dimensional analysis of M-estimators with decomposable regularizers. Stat. Sci. **27**(4), 538–557 (2012)
308. Y. Nesterov, *Introductory Lectures on Convex Optimization: A Basic Course* (Kluwer Academic Publishers, New York, 2004)
309. Y. Nesterov, Gradient methods for minimizing composite objective function. Tech. rept. Center for Operations Research and Econometrics (CORE), Catholic University of Louvain (UCL) (2007)
310. Y. Nesterov, A. Nemirovskii, *Interior-Point Polynomial Algorithms in Convex Programming* (SIAM, Philadelphia, 1994)
311. M.L. Overton, R.S. Womersley, On the sum of the largest eigenvalues of a symmetric matrix. SIAM J. Matrix Anal. Appl. **13**(1), 41–45 (1992)
312. G. Pisier, *The Volume of Convex Bodies and Banach Space Geometry* (Cambridge University Press, Cambridge, 1989)
313. D. Pollard, *Convergence of Stochastic Processes* (Springer, New York, 1984)
314. D. Pollard, *Empirical Processes: Theory and Applications*. NSF-CBMS Regional Conference Series in Probability and Statistics, vol. 2 (Institute of Mathematical Statistics, Hayward, 1990)
315. A. Prékopa, On logarithmic concave measures and functions. Acta Scientiarum Mathematicarum **34**, 335–343 (1973)
316. G. Raskutti, M.J. Wainwright, B. Yu, Restricted eigenvalue properties for correlated Gaussian designs. J. Mach. Learn. Res. **11**, 2241–2259 (2010)
317. G. Raskutti, M.J. Wainwright, B. Yu, Minimax rates of estimation for high-dimensional linear regression over ℓ_q-balls. IEEE Trans. Inform. Theory **57**(10), 6976–6994 (2011)
318. G. Raskutti, M.J. Wainwright, B. Yu, Minimax-optimal rates for sparse additive models over kernel classes via convex programming. J. Mach. Learn. Res. **13**, 389–427 (2012)
319. P. Ravikumar, M.J. Wainwright, J.D. Lafferty, High-dimensional Ising model selection using ℓ_1-regularized logistic regression. Ann. Stat. **38**(3), 1287–1319 (2010)
320. P. Ravikumar, M.J. Wainwright, G. Raskutti, B. Yu, High-dimensional covariance estimation by minimizing ℓ_1-penalized log-determinant divergence. Electron. J. Stat. **5**, 935–980 (2011)
321. B. Recht, A simpler approach to matrix completion. J. Mach. Learn. Res. **12**, 3413–3430 (2011)
322. B. Recht, M. Fazel, P. Parrilo, Guaranteed minimum rank solutions to linear matrix equations via nuclear norm minimization. SIAM Rev. **52**(3), 471–501 (2010)
323. G. Reinsel, R. Velu, *Multivariate Reduced-Rank Regression: Theory and Applications*. Lecture Notes in Statistics (Springer, New York, 1998)
324. P. Rigollet, Kullback-Leibler aggregation and misspecified generalized linear models. Ann. Stat. **40**(2), 639–665 (2012)

325. P. Rigollet, A.B. Tsybakov, Exponential screening and optimal rates of sparse estimation. Ann. Stat. **39**(2), 731–771 (2011)
326. H.P. Rosenthal, On the subspaces of $L^p(p > 2)$ spanned by sequences of independent random variables. Israel J. Math. **8**, 273–303 (1970)
327. M. Rudelson, Random vectors in the isotropic position. J. Funct. Anal. **164**, 60–72 (1999)
328. M. Rudelson, R. Vershynin, Sparse reconstruction by convex relaxation: Fourier and Gaussian measurements, in *Conference on Information Sciences and Systems (CISS)* (2006)
329. M. Rudelson, R. Vershynin, On sparse reconstruction from Fourier and Gaussian measurements. Commun. Pure Appl. Math. **61**(8), 1025–1045 (2008)
330. M. Rudelson, R. Vershynin, Hanson-Wright inequality and sub-Gaussian concentration. Electron. Commun. Probab. **18**(82), 1–9 (2013)
331. M. Rudelson, S. Zhou, Reconstruction from anisotropic random measurements. IEEE Trans. Inform. Theory **59**(6), 3434–3447 (2013)
332. D. Ruppert, M.P. Wand, R.J. Carroll, *Semiparametric Regression* (Cambridge University Press, Cambridge, 2003)
333. R.J. Samworth, Recent progress in log-concave density estimation. Stat. Sci. **33**(4), 493–509 (2018)
334. L.K. Saul, M.I. Jordan, Mixed memory Markov models: decomposing complex stochastic processes as mixtures of simpler ones. Mach. Learn. **37**, 75–87 (1999)
335. W.O. Saxton, W. Baumeister, The correlation averaging of a regularly arranged bacterial cell envelope protein. J. Microsc. **127**(2), 127–138 (1982)
336. I.J. Schoenberg, Metric spaces and completely monotone functions. Ann. Math. **39**, 811–841 (1938)
337. B. Schölkopf, A.J. Smola, *Learning with Kernels* (MIT Press, Cambridge, 2002)
338. B. Schölkopf, A.J. Smola, K.R. Müller, Nonlinear component analysis as a kernel eigenvalue problem. Neural Comput. **10**, 1299–1319 (1998)
339. G. Schwarz, Estimating the dimension of a model. Ann. Stat. **6**(2), 461–464 (1978)
340. C. Scott, R.D. Nowak, Minimax-optimal classification with dyadic decision trees. IEEE Trans. Inform. Theory **52**(4), 1335–1353 (2006)
341. B. Sen, M. Meyer, Testing against a linear regression model using ideas from shape-restricted estimation. J. Roy. Stat. Soc. B **79**(2), 423–448 (2017)
342. J. Shawe-Taylor, N. Cristianini, *Kernel Methods for Pattern Analysis* (Cambridge University Press, Cambridge, 2004)
343. A.J. Smola, B. Schölkopf, Sparse greedy matrix approximation for machine learning, in *Proceedings of the International Conference on Machine Learning* (2000), pp. 911–918
344. D. Sontag, T. Meltzer, A. Globerson, T. Jaakkola, Y. Weiss, Tightening LP relaxations for MAP using message passing, in *Proceedings of the Conference on Uncertainty in Artificial Intelligence (UAI)* (2008), pp. 503–510
345. P. Speckman, Kernel smoothing in partial linear models. J. Roy. Stat. Soc. B **50**, 413–436 (1988)
346. N. Srebro, T. Jaakkola, Weighted low rank approximation, in *Proceedings of the International Conference on Machine Learning* (2003), pp. 720–727
347. N. Srebro, J.D.M. Rennie, T.S. Jaakkola, Maximum-margin matrix factorization, in *Advances in Neural Information Processing Systems*, vol. 17 (2005), pp. 1329–1336
348. I. Steinwart, A. Christmann, *Support Vector Machines* (Springer, New York, 2008)
349. C.M. Stein, Estimation of the mean of a multivariate normal distribution. Ann. Stat. **9**(6), 1135–1151 (1981)
350. G.W. Stewart, J.G. Sun, *Matrix Perturbation Theory* (Academic Press, Boston, 1990)
351. M. Stone, Cross-validatory choice and assessment of statistical predictions. J. Roy. Stat. Soc. B **36**, 111–147 (1974)
352. C.J. Stone, Optimal global rates of convergence for nonparametric regression. Ann. Stat. **10**(4), 1040–1053 (1982)
353. M. Talagrand, Sharper bounds for Gaussian and empirical processes. Ann. Probab. **22**, 28–76 (1994)

354. M. Talagrand, Concentration of measure and isoperimetric inequalities in product spaces. Publications Mathématiques de l'IHES **81**, 73–205 (1995)
355. M. Talagrand, New concentration inequalities in product spaces. Inventiones Mathematicae **126**, 505–563 (1996)
356. M. Talagrand, *The Generic Chaining* (Springer, New York, 2005)
357. R. Tibshirani, Regression shrinkage and selection via the Lasso. J. Roy. Stat. Soc. B **58**(1), 267–288 (1996)
358. R. Tibshirani, M. Saunders, S. Rosset, J. Zhu, K. Knight, Sparsity and smoothness via the fused Lasso. J. Roy. Stat. Soc. B **67**(1), 91–108 (2005)
359. A.N. Tikhonov, Solution of incorrectly formulated problems and the regularization method. Soviet Math. Doklady **4**, 1035–1038 (1963)
360. K.C. Toh, M.J. Todd, R.H. Tütüncü, SDPT3—a MATLAB software package for semidefinite programming. Optim. Methods Softw. **11**, 545–581 (1999)
361. J.A. Tropp, Just relax: convex programming methods for identifying sparse signals. IEEE Trans. Inform. Theory **51**(3), 1030–1051 (2006)
362. J.A. Tropp, On the conditioning of random subdictionaries. Appl. Comput. Harmon. Anal. **25**, 1–24 (2008)
363. J.A. Tropp, User-friendly tail bounds for sums of random matrices. Found. Comput. Math. **12**(4), 389–434 (2012)
364. A.B. Tsybakov, Optimal aggregation of classifiers in statistical learning. Ann. Stat. **32**(1), 135–166 (2004)
365. A.B. Tsybakov, *Introduction to Nonparametric Estimation* (Springer, New York, 2009)
366. A.B. Tsybakov, S.A. van de Geer, Square root penalty: adaptation to the margin in classification and in edge estimation. Ann. Stat. **33**(3), 1203–1224 (2005)
367. L.G. Valiant, A theory of the learnable. Commun. ACM **27**(11), 1134–1142 (1984)
368. S.A. van de Geer, *Empirical Processes in M-Estimation* (Cambridge University Press, Cambridge, 2000)
369. S.A. van de Geer, P. Bühlmann, On the conditions used to prove oracle results for the Lasso. Electron. J. Stat. **3**, 1360–1392 (2009)
370. A.W. van der Vaart, *Asymptotic Statistics* (Cambridge University Press, Cambridge, 1998)
371. A.W. van der Vaart, J.A. Wellner, *Weak Convergence and Empirical Processes* (Springer, New York, 1996)
372. V.N. Vapnik, *Statistical Learning Theory* (Wiley, New York, 1998)
373. V.N. Vapnik, A.Y. Chervonenkis, On the uniform convergence of relative frequencies of events to their probabilities. Theory Probab. Appl. **16**(2), 264–280 (1971)
374. R. Vershynin, Introduction to the non-asymptotic analysis of random matrices, in *Compressed Sensing: Theory and Applications* (Cambridge University Press, Cambridge, 2012)
375. M.J. Wainwright, Sharp thresholds for high-dimensional and noisy sparsity recovery using ℓ_1-constrained quadratic programming (Lasso). IEEE Trans. Inform. Theory **55**(5), 2183–2202 (2009)
376. M.J. Wainwright, Information-theoretic limits on sparsity recovery in the high-dimensional and noisy setting. IEEE Trans. Inform. Theory **55**(12), 5728–5741 (2009)
377. M.J. Wainwright, Structured regularizers for high-dimensional problems: statistical and computational issues. Ann. Rev. Stat. Appl. **1**, 233–253 (2014)
378. M.J. Wainwright, M.I. Jordan, Graphical models, exponential families and variational inference. Found. Trends Mach. Learn. **1**(1–2), 1–305 (2008)
379. M.J. Wainwright, T.S. Jaakkola, A.S. Willsky, Tree-based reparameterization framework for analysis of sum-product and related algorithms. IEEE Trans. Inform. Theory **49**(5), 1120–1146 (2003)
380. M.J. Wainwright, T.S. Jaakkola, A.S. Willsky, Tree consistency and bounds on the performance of the max-product algorithm and its generalizations. Stat. Comput. **14**, 143–166 (2004)
381. M.J. Wainwright, E.P. Simoncelli, Scale mixtures of Gaussians and the statistics of natural images, in *Advances in Neural Information Processing Systems*, vol. 12 (2000), pp. 855–861

382. G. Wahba, *Spline Models for Observational Data*. CBMS-NSF Regional Conference Series in Applied Mathematics (SIAM, Philadelphia, 1990)
383. L. Wang, M.D. Gordon, J. Zhu, Regularized least absolute deviations regression and an efficient algorithm for parameter tuning, in *Proceedings of the IEEE International Conference on Data Mining (ICDM)* (2006), pp. 690–696
384. M. Wax, T. Kailath, Detection of signals by information theoretic criteria. IEEE Trans. Acoust. Speech Signal Process. **33**(2), 387–392 (1985)
385. Y. Weiss, W.T. Freeman, Correctness of belief propagation in Gaussian graphical models of arbitrary topology. Neural Comput. **13**(10), 2173–2200 (2001)
386. P. Whittle, On the fitting of multivariate autoregressions and the approximate canonical factorization of a spectral density matrix. Biometrika **50**(1,2), 129–134 (1963)
387. N. Wiener, *Extrapolation, Interpolation and Smoothing of Stationary Time Series* (MIT Press, Cambridge, 1949)
388. D.M. Witten, R. Tibshirani, T. Hastie, A penalized matrix decomposition, with applications to sparse principal components and canonical correlation analysis. Biostatistics **10**(3), 515–534 (2009)
389. W.B. Wu, M. Pourahmadi, Nonparametric estimation of large covariance matrices of longitudinal data. Biometrika **90**(4), 831–844 (2003)
390. Y. Wu, S. Verdu, Functional properties of minimum mean-square error and mutual information. IEEE Trans. Inform. Theory **58**(3), 1289–1301 (2012)
391. B. Yu, Assouad, Fano and Le Cam, in *Festschrift for Lucien Le Cam* (Springer, Berlin, 1997), pp. 423–435
392. M. Yuan, High dimensional inverse covariance matrix estimation via linear programming. J. Mach. Learn. Res. **11**, 2261–2286 (2010)
393. M. Yuan, Y. Lin, Model selection and estimation in regression with grouped variables. J. Roy. Stat. Soc. B **68**(1), 49–67 (2006)
394. M. Yuan, Y. Lin, Model selection and estimation in the Gaussian graphical model. Biometrika **94**(1), 19–35 (2007)
395. C.H. Zhang, Nearly unbiased variable selection under minimax concave penalty. Ann. Stat. 38(2), 894–942 (2010)
396. C.H. Zhang, J. Huang, The sparsity and bias of the Lasso selection in high-dimensional linear regression. Ann. Stat. **36**(4), 1567–1594 (2008)
397. C.H. Zhang, T. Zhang, A general theory of concave regularization for high-dimensional sparse estimation problems. Stat. Sci. **27**(4), 576–593 (2012)
398. P. Zhao, B. Yu, On model selection consistency of Lasso. J. Mach. Learn. Res. **7**, 2541–2563 (2006)
399. P. Zhao, G. Rocha, B. Yu, The composite absolute penalties family for grouped and hierarchical variable selection. Ann. Stat. **37**(6A), 3468–3497 (2009)
400. S. Zhou, Restricted eigenvalue conditions on subgaussian random matrices. Tech. rept. Department of Mathematics, University of Michigan (2009)
401. H. Zou, The adaptive Lasso and its oracle properties. J. Am. Stat. Assoc. **101**(476), 1418–1429 (2006)
402. H. Zou, T. Hastie, Regularization and variable selection via the elastic net. J. Roy. Stat. Soc. B **67**(2), 301–320 (2005)
403. H. Zou, T. Hastie, R. Tibshirani, Sparse principal component analysis. J. Comput. Graph. Stat. **15**(2), 265–286 (2006)
404. A. Zygmund, *Trigonometric Series* (Cambridge University Press, Cambridge, 2002)
405. T. Zhang, Statistical behavior and consistency of classification methods based on convex risk minimization. Ann. Stat. **32**(1), 56–85 (2004)
406. T. Zhang, Some sharp performance bounds for least squares regression with L1 regularization. Ann. Stat. **37**(5A), 2109–2144 (2009)
407. T. Zhang, Analysis of multi-stage convex relaxation for sparse regularization. J. Mach. Learn. Res. **11**, 1081–1107 (2010)

408. T. Zhang, Sparse recovery with orthogonal matching pursuit under RIP. IEEE Trans. Inform. Theory **57**(9), 6215–6221 (2011)
409. T. Zhang, F.J. Oles, Text categorization based on regularized linear classification methods. Inform. Retrieval **4**(1), 5–31 (2001)
410. P. Zhao, B. Yu, Boosted Lasso. Tech. rept. Statistics Department, UC Berkeley (2004)
411. S. Zhou, Thresholding procedures for high dimensional variable selection and statistical estimation, in *Advances in Neural Information Processing Systems*, vol. 22 (2009), pp. 2304–2313
412. H. Zou, The adaptive elastic-net procedure. J. Comput. Graph. Stat. **18**(4), 1048–1066 (2009)
413. H. Zou, T. Hastie, Addendum: regularization and variable selection via the elastic net. J. Roy. Stat. Soc. B **67**(5), 768 (2005)
414. H. Zou, H. H. Zhang, On the adaptive elastic-net with a diverging number of parameters. Ann. Stat. **37**(4), 1733–1751 (2009)
415. A. Zouzias, N. M. Freris, Randomized extended Kaczmarz for solving least squares. SIAM J. Matrix Anal. Appl. **34**(2), 773–793 (2013)
416. M. Zhu, T. J. Hastie, Feature extraction for nonparametric discriminant analysis. J. Comput. Graph. Stat. **12**(1), 101–120 (2003)

MIX
Papier aus verantwortungsvollen Quellen
Paper from responsible sources
FSC® C105338

If you have any concerns about our products,
you can contact us on
ProductSafety@springernature.com

In case Publisher is established outside the EU,
the EU authorized representative is:
Springer Nature Customer Service Center GmbH
Europaplatz 3, 69115 Heidelberg, Germany

Printed by Libri Plureos GmbH
in Hamburg, Germany